Advances in Biochemical Engineering 2

Edited by

T. K. Ghose, A. Fiechter, and
N. Blakebrough

With 70 Figures

Springer-Verlag Berlin Heidelberg GmbH 1972

T. K. Ghose

Dept. of Chemical Engineering, Indian Institute of Technology, New Delhi/India

A. Fiechter

Mikrobiologisches Institut der Eidgen. Techn. Hochschule, Zürich/Schweiz

N. Blakebrough

The University of Birmingham
Dept. of Chemical Engineering, Birmingham 15/Great Britain

ISBN 978-3-662-15586-8   ISBN 978-3-540-37990-4 (eBook)
DOI 10.1007/978-3-540-37990-4

Advances
in Biochemical
Engineering 2

Edited by

T. K. Ghose, A. Fiechter, and
N. Blakebrough

With 70 Figures

Springer-Verlag Berlin Heidelberg GmbH 1972

T. K. Ghose

Dept. of Chemical Engineering, Indian Institute of Technology, New Delhi/India

A. Fiechter

Mikrobiologisches Institut der Eidgen. Techn. Hochschule, Zürich/Schweiz

N. Blakebrough

The University of Birmingham
Dept. of Chemical Engineering, Birmingham 15/Great Britain

ISBN 978-3-662-15586-8   ISBN 978-3-540-37990-4 (eBook)
DOI 10.1007/978-3-540-37990-4

Contents

CHAPTER 1

Enzyme Engineering

LEMUEL B. WINGARD, JR. *

With 5 Figures

Contents

1. Introduction

This review is designed to acquaint the reader with the new area of specialization called "Enzyme Engineering", to describe the several sub-topics that make up this area, and to present the status of development

* Supported by Public Health Service NIGMS Special Fellowship No. 5F03GM44407.

of each of the sub-topics. It is anticipated that in-depth reviews of each of the sub-topics will be covered by other authors. This extensive, but intentionally incomplete, review of the literature extends through early 1972.

a) What is Enzyme Engineering?

Enzymes are the proteinaceous compounds that have the unique ability to catalyze the numerous reactions that occur in living organisms. The uniqueness of these compounds stems from the high degree of reactant (usually called substrate) specificity and "mild" reaction conditions under which enzyme catalysts operate. Some enzymes are specific for a particular type of chemical linkage, e.g. protease hydrolysis of the peptide bond, while others may be specific for a particular optical isomer of a compound e.g. the oxidation of β-D-glucose by glucose oxidase. In addition, the physiological conditions of temperature, pressure, pH, and concentration (enzyme and substrate) under which most enzymes function *in vivo* are "mild" when compared to the conditions required for "useful" reaction rates with most non-enzymatic catalysts.

For many years enzyme catalysts have been utilized widely in analytical chemistry, clinical medicine, and industrial processing. In general these applications have involved 1. crude mixtures of isolated enzymes or 2. fermentation processes where the enzymes either have remained an internal part of the microorganisms that carried out the fermentation or else have been secreted by the microorganisms into the fermentation broth. Until recently, little use has been made of highly purified enzymes except as tools in biochemical research.

For perhaps 25 years scientists and engineers have speculated on the application as well as the practicality of the controlled use of individual enzyme catalysts *in vitro* as well as *in vivo*. In recent years a number of developments have occurred to increase our knowledge of enzyme properties and mechanisms, to improve our ability to isolate and purify individual enzymes, and to facilitate the recovery of enzymes for re-use in a variety of reactor configurations. Therefore, the large number of potential uses for individual enzyme catalysts (often of high purity) in industry, analytical techniques, and medicine is moving now from speculation to reality; although in most cases practicality of the end result remains to be demonstrated. But this shift from speculation to reality has generated a strong interest in testing the practicality of the numerous end uses that have been envisaged. "Enzyme Engineering" is the name given to this activity. It includes the production, isolation, and purification of the desired enzyme(s); the development of a practical reactor and associated process for economical use and re-use of the enzyme(s); and the demonstration that the end result as done with enzyme catalysts has some useful advantage over obtaining the same end result by a non-enzymatic method. A more detailed discussion of the factors that brought about the development of this area called "Enzyme Engineering" appears elsewhere (Wingard, 1972a).

A book on "Enzyme Engineering" (Wingard, 1972b), an assessment of industrial applications of enzyme technology (Rubin, 1972), and a review of industrial uses of enzymes (De Becze, 1965) are available. The numerous reviews and books on specific sub-topics of "Enzyme Engineering" are listed in this review with the discussion of each sub-topic.

b) Definitions and Properties

Approximately 1500 enzymes have been found, and it is likely that many others exist (IUB, 1965; Barman, 1969). The six classes of enzymes, as defined by the International Union of Biochemistry, along with an example of each are shown in Fig. 1. The ligases also are referred to as

Oxidoreductases (oxidation-reduction)

$$Alcohol + NAD^+ \rightleftharpoons Aldehyde + NADH + H^+$$
$$(1.1.1.1)$$

Transferases (group transfer)

$$ATP + Pyruvate \rightleftharpoons ADP + Phosphoenolpyruvate$$
$$(2.7.1.40)$$

Hydrolases (hydrolysis)

$$Peptidyl\text{-}L\text{-}Amino\ Acid + H_2O \rightleftharpoons Peptide + L\text{-}Amino\ Acid$$
$$(3.4.2.1)$$

Lyases (remove or add on double bond)

$$Oxalate \rightleftharpoons Formate + CO_2$$
$$(4.1.1.2)$$

Isomerases (isomerization)

$$D\text{-}Mannose \rightleftharpoons D\text{-}Fructose$$
$$(5.3.1.7)$$

Ligases (join molecules, use high energy bonds)

$$ATP + Succinate + CoA \rightleftharpoons ADP + orthophosphate + Succinyl - CoA$$
$$(6.2.1.5)$$

Fig. 1. Classes of enzymes (IUB 1965). (Number) refers to International Union of Biochemistry identification number. (1.1.1.1) = alcohol:NAD oxidoreductase *or* alcohol dehydrogenase. (2.7.1.40) = ATP: pyruvate phosphotransferase *or* pyruvate kinase. (3.4.2.1) = peptidyl-L-amino-acid hydrolase *or* carboxypeptidase A. (4.1.1.2) = oxalate carboxy-lyase *or* oxalate decarboxylase. (5.3.1.7) = D-mannose ketol isomerase *or* mannose isomerase. (6.2.1.5) = succinate:CoA ligase (ADP) *or* succinyl-CoA synthetase

synthetases. Of course many enzymes are better known by their trivial or common names, but the use of such nomenclature should be minimized to avoid confusion. All papers dealing with specific enzymes should cite the International Union of Biochemistry identification numbers to specify exactly the enzymes being discussed.

Many enzymes require the presence of co-factors for appreciable catalytic activity. In some cases the co-factors normally are free in solution and appear to complex or interact with the enzyme protein or the substrate only during the reaction sequence of substrate going to product. Other co-factors normally are bound to the enzyme protein essentially all the time; however they can be removed by dialysis. Table 1 is a list of the more common co-factors. Suitable methods for supplying the co-factors to various enzyme reactor configurations likely will be the major problem for the practical *in vitro* utilization of certain enzymes. In the case of FAD and NAD the cost of the co-factor may exceed that of the enzyme unless appropriate means for recovery and re-use are available.

Table 1. Common co-factors for enzyme catalysis

Pyridoxal phosphate
Thiamine pyrophosphate
NAD (nicotinamide-adenine dinucleotide) (new name for DPN)
NADP (nicotinamide-adenine dinucleotide phosphate) (new name for TPN)
FAD (flavin-adenine dinucleotide)
FMN (flavin mononucleotide)
ATP, Mg^{2+}
Glutathione
Coenzyme A
Thiol esters

A number of techniques has been developed for holding an enzyme within the general confines of a reactor while continuously passing a solution of substrate into the reactor and drawing off a product/substrate solution. In keeping with the recommended nomenclature, the term "immobilized" will be used in this review to describe all such processes for the retention of enzymes. Fig. 2 is a description of the nomenclature scheme recommended by a group of scientists and engineers active in research with immobilized enzymes (Sundaram and Pye, 1972). With immobilized enzymes one needs to place special emphasis on the planning and complete reporting of the conditions used for measuring the activity. Since the purpose of an enzyme is to catalyze a reaction, the activity of a specific enzyme preparation is defined in terms of the rate of reaction produced by a given amount of protein under a specified set of

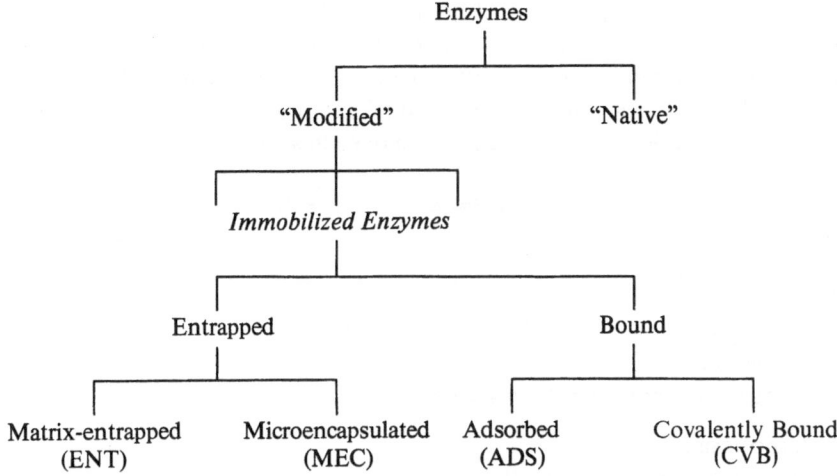

Fig. 2. Nomenclature for immobilized enzymes. Initials in parentheses go between the support and the enzyme, as cellulose-CVB-trypsin; ref. Sundarum and Pye (1972)

test conditions. With immobilized enzymes often it is not possible to assay for enzyme activity without the presence of significant diffusional resistances. In such cases the measured rate is a combination of the rates due to chemical reaction and diffusion. Therefore, with immobilized enzymes especially it is important that the detailed conditions for the assay be reported. Some guidelines as recommended by a group of persons active in research with immobilized enzymes are as follows (Sundaram and Pye, 1972): "The activity of an immobilized enzyme should be reported as an initial reaction rate (moles/min) per mg of the dry protein preparation, measured under clearly specified conditions. Where enzymes are bound to surfaces (tubes, plates, membranes, etc.) the activity may be reported as the initial reaction rate (moles/min) per unit area of covered surface under clearly specified conditions. The drying conditions of the enzyme preparation should always be specified, and whenever possible the protein content of the dry preparation should be reported". This group also goes on to recommend that the experimentally determined values for the maximum reaction velocity and the Michaelis constant of immobilized enzymes be referred to as "apparent" values.

Unless otherwise stated, the quantitative description of the rate of an enzyme-catalyzed reaction, as used in this review, will follow the approach first developed by Henri (1903), clarified by Michaelis and Menten (1913), and modified to the presently used form by Briggs and Haldane (1925). The main points of the Briggs-Haldane approach are noted here for later use and for emphasis of the underlying assumptions. The presence of enzyme substrate intermediates or complexes is fairly well accepted. Bell and Koshland (1971) list 60 examples where there is strong experimental evidence for the existence of such species.

$$S + E \underset{k_2}{\overset{k_1}{\rightleftharpoons}} ES \xrightarrow{k_3} P + E$$

where:

S = substrate
E = enzyme
ES = enzyme substrate complex
P = product
k_n = rate constant for n-th reaction

Assumptions:

a) Reaction 3 is rate-controlling so that

$$V_m = k_3 [E]_0 .$$

b) $d[ES]/dt = 0$.
c) $[E]_0 = [E] + [ES]$.
d) $[S]_0 \gg [E]_0$ and $[S] \approx [S]_0$.

With these assumptions, the equation for the *initial* reaction velocity v_0 becomes:

$$\frac{d[S]_0}{dt} = v_0 = \frac{V_m[S]_0}{K_m + [S]_0} . \tag{1}$$

$[\ ]$ = concentration
$[\ ]_0$ = initial concentration
v = reaction velocity, dS/dt
V_m = maximum reaction velocity
K_m = Michaelis constant = $\dfrac{k_2 + k_3}{k_1}$

Assumption "*d*" often is overlooked; and while the results may still be sufficiently accurate for many engineering purposes, one should be aware of potential errors due to overlooking the relationship between the initial substrate and enzyme concentrations as well as the extent of reaction of the substrate.

Enzymes possess various degrees of stability to environmental factors, such as ionic strength, temperature, and pH. In most cases, enzyme activity is appreciable only over a relatively narrow span of environmental conditions. Exposure of enzymes to extremes of temperature or ionic strength or to certain additives causes denaturation and loss of enzyme activity. With the high initial cost of many enzymes, there is considerable economic incentive for the development of methods for renaturation of inactive or denatured enzymes. Some success in the renaturation and return of activity of denatured enzymes has been reported. Teipel and Koshland (1971) have followed the regain in chain folding and enzyme activity upon renaturation (by dilution of the denaturing agent) of fumarase, lactic dehydrogenase, aldolase, and other enzymes denatured by 6 M guanidine hydrochloride. Weber and Kuter (1971) and McKenzie and Ralston (1971) also have had some success in the renaturation of enzymes.

Since enzymes are proteins, immunological sensitization and dermatitis might be expected upon contact of humans with enzymes foreign to the body. In the case of the enzyme-containing detergents the enzymes were supplied as dust-producing powders which underwent airborne transport to the lungs of humans. Belin *et al.* (1970) and Lichtenstein *et al.* (1971) claim conclusive evidence for sensitization and respiratory disease in humans resulting from contact with enzyme-containing detergents. In a related safety consideration, Dubons (1971) describes potentially toxic substances found in detergent enzyme produced from autolysates of *B. subtilis* (the primary source for detergent enzymes). Perhaps as a result of the detergent enzyme-induced diseases, industry has learned that enzymes cannot be used with the same abandon as inorganic chemicals. In future industrial applications of dried enzyme preparations it may be necessary to keep the enzymes encapsulated prior to use or to take other precautionary measures. However, with careful planning there is no reason to doubt that adequate safety measures can be taken and enzymes used safely in industrial processing.

For the design and development of enzyme-containing reactors one will probably want to know numerical values for certain physical parameters, such as diffusion coefficients or viscosity. Numerous standard techniques for the experimental determination of these values are described in several texts on protein chemistry. Only a recently developed method is mentioned here in that it affords a means for obtaining sedimentation and diffusion coefficients for the active enzyme/substrate complex. Such data could be useful for some but not all enzyme-reactor systems. Cohen and Mire (1971) have described the method, which utilizes an ultracentrifuge, and have presented data for several enzyme/substrate systems.

2. Sources of Enzymes

Any living organism in theory is a source from which to obtain enzymes for industrial, analytical, and medical applications. However, all known enzymes do not appear to be present in all living organisms, at least not in recoverable or even detectable concentrations. Therefore, the task is one of finding an economical source that produces the desired enzyme in sufficient concentration and from which the enzyme can be isolated and often purified with minimal loss of activity. Microbial cells at present are the most widely used and versatile source of enzymes. In addition, this source is easily amenable to genetic manipulations thus providing an opportunity for the development of enzymes with more desirable operational parameters, such as pH optimum. Although plant and animal tissues are the major sources for several commercial enzymes, the longer growth cycle and the usually more difficult task of enzyme isolation often make these latter sources less economical than microbial cells. However, at present only a single source, be it microbial, animal, or plant, is known for many enzymes.

An alternate, but more futuristic, method for obtaining enzymes is through the chemical synthesis of enzymes or enzyme analogs. The latter are synthetic compounds, modeled after enzyme fragments, with retention of part of the activity but probably less of the specificity of native enzymes. And finally, the chemical modification of native enzymes, either before or after isolation and purification, potentially is a route to enzymes having more desirable operational parameters, especially for industrial processes.

a) Microbial, Plant, and Animal Sources of Enzymes

The field of microbiology is filled with papers and reviews on the production of enzymes from microbial cells. A few of the more pertinent papers are cited here. However, the main objective of this section is to describe some of the manipulations being done to favor the formation of specific enzymes or to produce enzymes with modified properties. The reader is directed to the literature of microbiology for more thorough coverage.

Enzymes can be divided into two major categories based on the *in vivo* location where they normally carry out their biological function. The extracellular or exo-enzymes are synthesized within the cell and then secreted into the extracellular space, whereas the intracellular enzymes are both synthesized and utilized entirely within the cell. In most cases the intracellular type either are present as aggregates or particles or are bound on or trapped within sub-cellular particles or intracellular membranes, thus making the isolation a much more difficult task. The majority of the commercially utilized enzymes are of the extracellular type. Lampen (1972) has pointed out that plant, animal, and fungal cells (eukaryotic cells) usually secrete large glycoprotein enzymes containing sulfur bridges, while bacteria secrete small protein enzymes free of sulfur bridges. Lampen (1972) and Schramm (1967) both describe the general process of how the enzymes get from the site of synthesis, through the endoplasmic reticulum, and into vesicles or granules which are expelled through the cell membrane into the extracellular space.

It is especially with the intracellular enzymes that much of the potential lies for the expanded *in vitro* use of enzymes. The extracellular enzymes in general tend to catalyze hydrolytic or degradative types of reactions; while the intracellular enzymes are those most associated with synthesis reactions. Many intracellular enzymes are present as highly organized multienzyme systems that catalyze an ordered sequence of reactions. Reed and Cox (1970) describe several such systems for the synthesis of fatty acids and aromatic amino-acids and for two oxidative decarboxylations. Other multienzyme systems are reviewed by Ginsburg and Stadtman (1970).

Irrespective of whether one is interested in producing extra- or intracellular enzymes, there are several types of manipulations that can be utilized to favor the production of the desired enzyme. The amount of enzyme produced by a cell results from a combination of genetic and enzymatic phenomena. Pardee (1969), with more recent examples by Demain (1972), has described several methods for increased formation of a particular enzyme: 1. choice of the proper strain and culture of microorganism; 2. nutrient additives for addition or removal of inducers or repressors of RNA-polymerase function; 3. choice of processing conditions, such as pH, temperature, light, oxygen transfer, media concentration, trace materials, and stage of growth cycle; 4. damage to cells so they cannot produce the repressor that decreases the functioning of RNA-polymerase (such cells are called constitutive mutants in that they can produce as much enzyme as completely induced or derepressed cells); 5. increase the number of structural genes for making the desired enzyme (called gene copies); and 6. increase the maximum rate of enzyme production of a constitutive mutant by making still another mutant. An additional review of the mechanism for control of gene expression is by Epstein and Beckwith (1968); and a number of industrial examples relating theory to practice are given by Elander (1969). In addition the induction of specific enzymes in microbial systems often can be aided by use of the substrate or its precursor for the desired enzyme as the carbon source in the nutrient mixture. Koch and Coffman (1970) have discussed the significance of diffusional resistances in determining the kinetics of enzyme induction using exogenous substrate.

Although peroxidase, urease, and other enzymes are obtained in commercial quantities mainly from plant tissue, much less work has been done with plant sources for enzymes as compared to microbial sources. Filner *et al.* (1969) describe the techniques and list over ninety reported cases of enzyme induction in plants brought about by a change in some environmental parameter. A later review by Marcus (1971) describes the induction of the cellulases by ethylene or indolylacetic acid, amylase by gibberellic acid, phenylalanine-ammonia lyase by phytochrome, and nitrate reductase by substrate addition. Both of the latter reviewers suggest that a major difference between plants and microbial cells may be the presence in plants of controlled degradation of enzymes as an added method of regulation of enzyme levels.

The majority of enzymes of animal origin are obtained from various organs, glands and to some extent blood. Because of the long time for growth and the difficulty in isolation, animals serve as a source of enzymes only under special conditions, such as there being no other source or local availability of a large supply of animal organs. Enzyme induction operates in animals as well as in plants and microbial systems and has been studied extensively in connection with the metabolism of drugs by liver enzymes. The subject of enzyme levels in animals has been reviewed by Schimke and Doyle (1970). The culture of animal tissue cells could be a unique source for many enzymes. Telling and Radlett (1970) have described the methodology for large-scale tissue culture, mainly as a means of virus production. Little emphasis apparently has been placed on the use of this technique for the production of intracellular enzymes.

b) Synthetic Enzymes, Analogs, and Modifications to Native Enzymes

Although the present sources of enzymes all depend on living cells to carry out the synthesis, there is a growing possibility that some future enzymes or enzyme fragments may be synthesized commercially by purely chemical methods. Just how long such developments will take to mature or how favorable the resulting economics will be remains to be seen.

Only one enzyme, ribonuclease, has been synthesized by chemical means, both by Denkewalter et al. (1969) and by Gutte and Merrifield (1969). Of particular importance has been the development of the method and automated equipment for solid-phase peptide synthesis. The latter method was followed by Gutte and Merrifield in their synthesis of the 1,2,4-aminoacid/ribonuclease A. The technique, developed by Merrifield, is described by Merrifield (1963) and Steward and Young (1969). The automated equipment is depicted by Merrifield et al. (1966). Marglin and Merrifield (1970) have reviewed the chemical synthesis of enzymes. Sano et al. (1971) have described the use of low cross-linked Amberlite XE 305 as the support material to decrease diffusional resistance in the solid-phase peptide method.

Another approach to the development of sources for enzyme catalysts is the synthesis of compounds that are easier to prepare than an enzyme yet possess some of the catalytic activity of the enzyme. Such compounds have been called enzyme analogs or enzyme models. The literature describing this approach is scattered through organic, inorganic, bio- and polymer chemistry. Only a few typical studies are cited here. Wingard and Finn (1966) have described some early attempts to prepare analogs of metal-containing enzymes. A more extensive review of enzyme analogs has been given by Lindsey (1969). Katchalski et al. (1964) have reviewed the catalytic activity of polyaminoacids and Yamamoto and Noguchi (1970) have demonstrated hydrolase activity with aminoacid copolymers. Palit (1968) has reviewed a number of vinyl polymers, lysine/glutamic acid copolymers, and surfactants showing enzyme-like activity. Sheehan (1967) has described the activity obtained on synthesizing peptides to simulate the active sites of specific enzymes.

Since histidine is found at the active site of many enzymes, much of the work with enzyme models has been done with compounds containing the imidazole group. Two studies which have demonstrated enzyme-like catalytic activity (greater than simple acid or base catalysis) for the hydrolysis of various esters have been reported by Overberger et al. (1967) and by Kunitake and Shinkai (1971). Klotz et al. (1971) have coined the term "Synzymes" to describe their enzyme-like hydrolytic polymers prepared from polyethyleneimines. One of the more classical studies in enzyme models was that of Wang (1955) in which a complex of ferric ion, hydroxyl ion, and triethylenetetramine gave a kinetic rate constant intermediate between that for catalase and the best inorganic catalyst for hydrogen peroxide decomposition. Additional reviews, citing work in Eastern Europe and Russia on enzyme models, are by Nikolaev (1964) and Khidekel (1968).

Still another approach for obtaining enzymes with the desired operational parameters, such as optimum pH and stability, involves the chemical modification of the active site or other important functional regions of native proteins. The use of the automated solid-phase peptide-synthesis technique may make it practical to prepare enzyme fragments possessing catalytic activity or even complete enzymes already having modified functional groups. On the other hand it may be more practical with certain enzymes to modify properties by chemically treating the natural enzyme. Most of these studies to date have been aimed at learning more about the mechanism of enzyme action. The work of Meighen et al. (1970) on the modification of aldolase, glyceraldehyde-3-phosphate dehydrogenase, and aspartate transcarbamylase by succinylation gave almost total loss of activity; while Stesina et al. (1971) claim to have modified the properties of mono- and diamine oxidase while retaining enzymatic activity. As will be discussed later, the proper choice of immobilization support material may also provide the desired modification of properties. Zaborsky (1972) has mentioned one of the few studies aimed specifically at the chemical modification of an enzyme to produce a desired change in an operational parameter.

3. Isolation and Purification of Enzymes

The majority of the present industrial applications of enzymes utilizes crude mixtures of enzymes or simple digests or extracts of microbial cells. However, much of the potential for industrial as well as analytical and medical utilization of enzymes lies in the controlled use of specific enzymes. In order to realize this potential relatively pure enzymes will be needed for many of the applications so that only the desired reactions are catalyzed and in proper sequence. Thus, economically sound techniques for moderate to large-scale isolation and purification of enzymes with little to no loss in enzyme activity take on special importance. Because of the ease by which most enzymes undergo loss of catalytic activity, many of the more common large-scale separation methods cannot be used. In general, one is limited to low-shear mechanical processes, solubility or partitioning processes, and a variety of processes based on the formation of weak attractive forces between the enzyme and a second phase. Thus ultrafiltration, extraction, zonal centrifugation, fractional precipitation, and dialysis are the methods most applicable for the initial isolation of enzymes from cellular debris and sub-cellular particles. Chromatography, gel filtration, extraction, ultracentrifugation, electrophoresis, ion exchange, affinity techniques, and ultrafiltration are most applicable for the selective fractionation and purification of an enzyme. Sober *et al.* (1964) have described the fractionation of proteins using these plus other methods, mainly of use in small-scale processing. A book related solely to enzyme purification and related techniques has been edited recently by Jakoby (1971). The latter gives fairly extensive descriptions of most of the above methods. In another review Edwards (1969) discusses the methodology for recovering biochemicals from fermentation broths, mainly emphasizing solids removal, ion exchange, gel filtration, and membrane methods. Dunnill and Lilly (1967b) have discussed the problems of equipment suitability, mixing, separation, chromatography, gel filtration, and electrodecantation as they relate to the isolation of enzymes on a large scale. In a later paper, these same authors (1972) describe some of their efforts and discuss many practical problems for isolating and purifying enzymes on a continuous basis. In one example they obtained a fifty to sixty times higher yield and sixteen times higher purity with continuous rather than batch isolation of the highly labile aminoacyl-tRNA-synthetase from a plant source.

a) Isolation of Crude Mixtures of Enzymes

The degree of difficulty in isolating crude mixtures of enzymes varies considerably with the source and the type of enzyme. For example, the task of isolation of extracellular enzymes from a fermentation broth is many times simpler than the isolation of intracellular particle-or membranebound enzymes. Extracellular enzymes can be isolated easily by filtration or centrifugation to remove cells and suspended solids, followed by washing or preparative-scale gel filtration and ultrafiltration for removal of fermentation nutrients and concentration of the normally dilute enzyme solutions.

For the isolation of intracellular enzymes the first step after collecting the cells involves breaking open the cell membrane, and in some cases also disrupting the membranes of specific sub-cellular particles. These membrane-rupture processes normally are done by mechanical, osmotic, chemical, or even enzymatic methods. Extraction and crude fractionation using inorganic salts (e.g. ammonium sulfate) or organic solvents is a common second step. Here it usually is necessary to concentrate the dilute enzyme solution, most often by ultrafiltration, for obtaining practicable levels of activity. The release of membrane-trapped or membrane-bound enzymes often presents special problems, as seen from the large number of special procedures appearing in the biochemical literature. In general, the same methods used for the disruption of cellular and sub-cellular membranes are employed to release membrane-trapped enzymes. Ultra-sonic devices and chemical agents, such as surfactants and various organic solvents, have proven highly useful. However, care must be used not to cause irreversible chain unfolding or shear degradation. The most difficult enzymes to isolate are those associated in a lipoprotein complex. Detergents, bile salts, and other chemical agents have been found very useful in solubilizing these enzymes in aqueous media. Since the true nature of lipoprotein enzymes and their often highly structured arrangements are not well understood, it is difficult to develop a rationale for the separation and purification of such enzymes. The role of the phospholipid part, often needed for enzymatic activity, remains to be clarified.

Dunnill et al. (1967a) have provided an example of the procedure and difficulties in the large-scale isolation of a crude enzyme, prolyl-tRNA-synthetase from mung bean. This highly labile enzyme was obtained with 10.5% recovery (42 g) from 25 kg of mung bean using a 22-step process.

Winchester *et al.* (1971) have provided a recent evaluation of electrode-cantation as a method for the isolation of enzymes, as an alternate to ammonium sulfate fractionation. The authors obtained 6 to 10 times enrichment with β-D-glucosidase from pig kidney as well as for several other enzymes but concluded that the method caused loss of enzyme activity with dilute solutions. Foster, Dunnill, and Lilly (1971) have studied the kinetics of salting-out with ammonium sulfate and have concluded that the following equation first developed for blood proteins also holds for enzyme proteins.:

$$\mathrm{Log} N = \beta - K C_s \qquad (2)$$

where: N is protein solubility, C_s is salt concentration, and both β and K are constants for each protein-salt system. Schwencke and coworkers (1971) have demonstrated the use of osmotic shock to release extracellular but not intracellular enzymes from yeast cells with only 20% loss of cell viability.

Charm and Lai (1971) have evaluated four methods for reducing material buildup at ultrafiltration membranes, a phenomena which slows the throughput, for ultrafiltration with protein micelles, cell debris, and catalase solution. The vibrating porous plate and vibrating plate filter were most effective for larger particles, while the design that gave laminar flow across the membrane and high wall-shear was best for dilute suspensions and proteins in solution. The fourth type with turbulent cross-flow showed no special attributes. The principles of ultrafiltration have been discussed by Michaels (1968). Porter (1972) has described the use of ultrafiltration specifically for the isolation and purification of enzymes and Goldsmith (1971) for macromolecules. Kozinski and Lightfoot (1971) have analyzed the effects of viscosity and diffusivity variations on concentration polarization during ultrafiltration of protein solutions in two-dimensional stagnation flow and found the effects to be small.

b) Purification of Enzymes

Although the optimum isolation and purification sequence and methodology for a particular enzyme will depend on the source, economics, scale of operation and enzyme stability, the techniques of adsorption or partition chromatography, gel filtration, and affinity methods usually are reserved for the final purification steps.

Many intracellular enzymes undergo considerable loss of activity during their isolation and purification. In many cases speed of operation is of utmost importance. In these cases, continuous processing may be faster and give less loss of activity as compared to batch processing. The previously mentioned example by Dunnill and Lilly (1972) points up the type of benefits to be obtained by using continuous processing for the isolation and purification of a highly labile enzyme such as aminoacyl-tRNA synthetase.

However, since chromatographic methods play a prime role in the purification steps, there is a need to reduce the theory of continuous chromatography to a more practical basis. Dunnill and Lilly (1972) describe briefly a continuous chromatography unit developed in their laboratory. Ito and Bowman (1971) have described their continuous-helix, countercurrent chromatographic unit for either small- or preparative-scale operation.

Gel filtration (also called gel-permeation chromatography, exclusion chromatography and gel chromatography) separates molecules on the basis of their molecular size. Filtration media include the popular cross-linked dextran gels (Sephadex), polyacrylamide gels (Bio-Gel) and the more recent agar or agarose gels (Sepharose, Bio-Gel). The technique is especially applicable to the fractionation of enzyme mixtures. Bombaugh (1970) has described the interrelationship of resolution, speed, and column loading on column performance and the repetitive use of a gel column for a given separation. Swanljung (1971) has described the use of a gradient of detergent concentration for the gel filtration purification of membrane enzymes. Detergents are used to solubilize membrane enzymes and allow the enzyme and membrane components to become separated partway through the gel column. However, as the separation proceeds and with the same detergent concentration throughout the column, the ratio of detergent to enzyme may become excessive and result in modification of enzyme properties. The use of a gradient in detergent concentration can prevent such modification and possible denaturation. Charm, Matteo, and Carlson (1969) have pointed up the need to maintain the ratio of column length to diameter as well as the Reynolds number constant in scale-up of gel filtration, assuming the same degree of packing uniformity and flow channeling in both large and small diameter columns.

A description of some recent experimental developments in the purification of enzymes by gel filtration and by affinity chromatography has been given by Porath (1972). Samples of 100 to 250 g of protein have been fractionated effectively on their largest column for low molecular weight proteins. The column measured 50 cm high by 90 cm diameter and contained about 300 l of Sephadex gel. The technique of affinity chromatography is a relatively new method based on the highly selective binding exhibited by enzymes, antibodies, and other proteins. Porath

(1972) has described recent developments in agarose support materials and has presented several examples utilizing these supports for affinity chromatography.

The general technique of affinity chromatography for enzymes is carried out by binding a specific ligand, such as a reversible enzyme inhibitor, to a water-insoluble carrier (i.e. agarose). Upon passing a mixture of crude enzymes through a column of the above material, the enzyme for which the inhibitor is specific is complexed and retained in the column. Subsequent passage of an eluting medium through the column displaces the retained enzyme. The general method of affinity chromatography has been reviewed by Cuatrecasas (1971) and Cuatrecasas and Anfinson (1971). The latter authors have listed over sixteen enzymes that have been separated by affinity chromatography. Some additional reports of enzyme purification by affinity chromatography are: trypsins by Robinson et al. (1971a), peptides with modified residues by Wilchek et al. (1971), dihydrofolate reductase by Newbold and Harding (1971) and by Kaufman and Pierce (1971), transaminases by Collier and Kohlhaw (1971) and proteolytic enzymes by Uren (1971). Weibel et al. (1972) and Lowe and Dean (1971) have reported using affinity chromatography with immobilized co-factors such as NAD. In some cases affinity chromatography has been used to isolate high-purity enzymes directly from a crude enzyme isolate thus bypassing a number of intermediate steps. Affinity methods appear to hold a promising future for the speedy, low-cost purification of enzymes; however, considerable development work remains to be carried out.

Another separation method often used for enzyme purification is electrophoresis. However, it is used mainly for analytical purposes in ascertaining the degree of purity or homogeneity of a purified enzyme. It finds little use so far in the preparative-scale purification of enzymes. The latter may change, though, if the theory for design of continuous electrophoretic equipment described by Vermeulen et al. (1971) can be put into practical hardware. The special method of polyacrylamide gel electrophoresis, also used mainly for analytical measurements with enzymes and so far used only with column loadings up to about a few grams, has been reviewed recently by Chrambach and Rodbard (1971).

4. Enzyme Immobilization

During the past several years a number of methods have been developed for retaining an enzyme within a limited region of space. Such methods fall under the general classification of "enzyme immobilization", although a number of other terms such as matrix-entrapped, insolubilized, encapsulated and support-coupled enzymes have been used. Several reasons are apparent for developing methods to immobilize enzymes. Through immobilization, enzyme recovery can be made easier and the resulting economics for *in vitro* applications at least in theory may become highly favorable. A second reason for wanting to immobilize enzymes is to provide an additional tool for investigating the mechanisms of enzyme activity and the factors that affect stability. And, finally, since most intracellular enzymes are attached to membranes or appear to be associated with interfaces, the study of immobilized enzymes may be a tool to improve understanding of the mode of action of intracellular, membrane-bound enzymes as well as a guide for developing novel *in vitro* devices for enzyme-controlled processing.

a) Methods

The general methods for immobilization of enzymes are shown schematically in Fig. 3. Goldman, Goldstein, and Katchalski (1971) have reviewed the methods and literature on the immobilization of enzymes through most of 1970. They list 16 enzymes immobilized by entrapment in polyacrylamide or starch gel (32 reported cases), 17 immobilized by

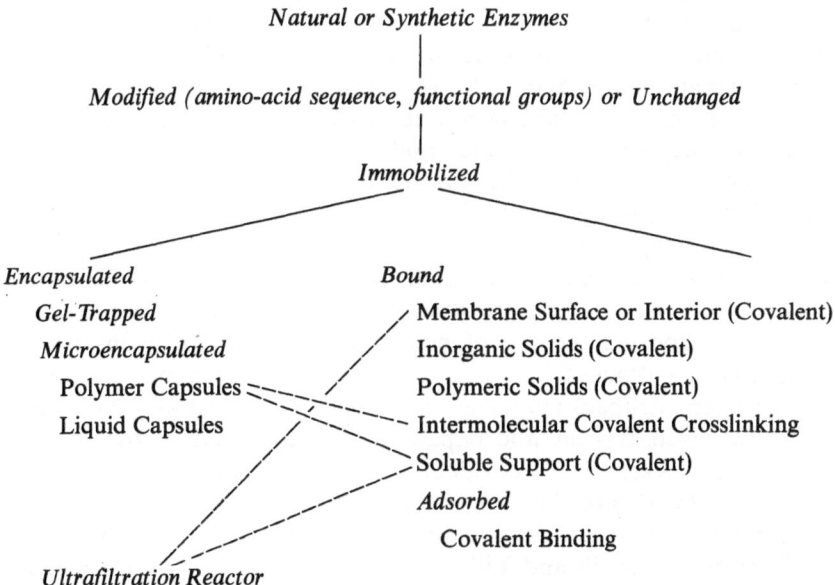

Fig. 3. Methods and some combinations for immobilization of enzymes

adsorption on various cellulosic and other materials (38 reported cases), 42 immobilized by covalent binding to cellulosic and various synthetic polymers and to porous glass (143 reported cases), and 7 enzymes immobilized by intermolecular crosslinking by glutaraldehyde and other disfunctional reagents (14 reported cases). Both the adsorption and gel entrapment methods suffer in that the former may lead to enzyme denaturation while the latter often imposes severe diffusional resistances to transport of large substrate and product molecules. Both adsorption and gel entrapment in general provide only temporary immobilization owing to reversible desorption and to leakage of enzyme from the gel matrix. Covalent bonding methods lead to permanent attachment of the enzyme to a support; however, many enzymes undergo loss of activity during the coupling process. In another recent review, Melrose (1971) has given a fairly complete coverage of the literature on the immobilization of enzymes (through 1970). He provides details of 178 reported cases of enzyme immobilization involving 54 different enzymes. Another review is by Goldstein (1969).

In addition to gel entrapment, adsorption, and covalent bonding (including cross-linkage) enzymes have been immobilized through encapsulation by semi-permeable membranes. Microencapsulation of enzymes to give collodion- or nylon-coated emulsified enzyme particles 1—100 microns in diameter, with the enzyme trapped inside the capsule, has been developed and studied by Chang (1964, 1972). The same concept of immobilization by encapsulation can be carried to larger and larger capsules and even to an industrial-size flow-reactor with semi-permeable membranes at the inlet and outlet and the enzyme trapped within. Most encapsulated enzymes have been free in solution; however, with high-molecular-weight substrates and moderate membrane permeability it may be necessary to attach the enzyme to a water-soluble polymer or to immobilize the enzyme to prevent its transport through the membrane and out of the capsule or reactor.

A few additional examples of enzyme immobilization, reported after the earlier review papers were prepared are described herein. Several authors have reported the use of new support materials. Monsan and Durand (1971) attached invertase to bentonite clay *via* cyanuric chloride; Axen and Ernback (1971) reported a mild treatment for attachment of chymotrypsin, trypsin, and papain to Sephadex and agarose beads; Vretblad and Axen (1971) covalently coupled pepsin to agarose; and Epton et al. (1971) attached several hydrolytic enzymes to cross-linked polyacryloylamino-acetaldehyde dimethylacetal with little loss in activity. Robinson, Dunnill and Lilly (1971) bound chymotrypsion and β-galactosidase with glutaraldehyde and Line et al. (1971) bound pepsin to porous glass with 50, 36, and 65% retention of activity, respectively.

Weibel, Weetall, and Bright (1971) and Larsson and Mosbach (1971) reported the immobilization of NAD(H) cofactor. Gregoriadis *et al.* (1971) described the encapsulation of amyloglucosidase in liposomes, a type of lipid spherule. Marshall, Walter, and Falb (1972) have immobilized hexokinase by adsorption on colloidal silica followed by covalent coupling to a cellulose derivative.

A group of persons active in enzyme immobilization studies (Sundaram and Pye, 1972) have suggested that the method of immobilization be included when cescribing an immobilized enzyme. The suggested system of nomenclature is described in Fig. 2. For example, trypsin covalently bound to carboxymethylcellulose would be written as "CM-cellulose CVB trypsin" and nylon microencapsulated urease written as "nylon MEC urease".

b) Properties of Immobilized Enzymes

The stability of an enzyme to temperature, pH, ionic strength, substrate and enzyme concentrations, trace materials, and other factors is of prime interest for any contemplated process or device utilizing enzyme catalysis. Likewise the usable kinetic activity and the conditions for maximizing this activity are equally important. Most of our knowledge about enzyme stability and kinetics has been obtained with free enzymes in solution. However, the recent efforts to immobilize enzymes have included a number of studies on the stability and kinetic activity of immobilized enzymes. Generalizations based on the scattered, and sometimes conflicting, data available to date are rather difficult to make. However, it appears that a moderate number of immobilized enzymes are more stable under certain conditions than are the same enzymes free in solution. In many cases the stability, kinetic activity, and other parameters are modified for immobilized *versus* free enzymes; with modifications in either direction. In general the detailed mechanism for a change in a specific property or parameter of a particular enzyme upon immobilization is not known; however numerous qualitative reasons can be suggested which in a number of cases are in agreement with experimental data. A second generality regarding many immobilized enzyme systems is the extensive diffusional resistance for transport of substrates and product to and from the active site. This resistance often becomes rate-limiting for many immobilized reaction systems. The subject of diffusion will be dealt with in more detail in the section on "Kinetics and Reactors".

Some of the factors that strongly influence the type and degree of property/activity, property/stability modification upon immobilization of enzymes are: 1. *Nature of the Support Material*; especially its hydrophobic/hydrophilic character, localized electrostatic field strengths, and spatial requirements as related to support/enzyme steric interactions and diffusional resistances.

2. *Physico-chemical Severity of Immobilization Procedure;* free radical polymerization reactions in the presence of the enzyme as well as covalent coupling reactions with enzyme molecules may cause conformational changes or functional group modification leading to decreased kinetic activity or inactivation.

3. *Enzyme Site for Immobilization;* although the type of functional group to be used for the coupling process usually can be selected, the specific aminoacid residue to be used on the enzyme is beyondour control at present; in some cases it may be desirable to protect certain functional groups or to protect the active site with a reversible inhibitor during the immobilization procedure to retain high kinetic activity.

4. *Interfacial Orientation of Enzyme;* enzymes bound to membranes *in vivo* may show high kinetic activity only at interfaces such as those formed by proteo-lipid structures so that *in vitro* applications may need to simulate such structures.

The effects of these above factors on the properties and kinetic activity of immobilized enzymes has been reviewed recently by several authors. Noted here are a few of their major conclusions along with some more recent developments. Goldman, Goldstein, and Katchalski (1971) concluded that the stability of proteolytic enzymes could be improved by immobilization owing to reduction in autodigestion, and that adsorbed or covalently bound enzymes usually showed decreased thermal stability. These authors presented a lengthy discussion of the effect of electrostatic interactions, pH, and ionic strength on kinetic parameters with special reference to membrane-bound enzymes. In a similar review, Katchalski, Silman, and Goldman (1971) discussed the effect of the microenvironment of membrane-bound enzymes on their possible mode of action both *in vivo* and *in vitro* with special emphasis on the role of the hydrophobic lipid component. Melrose (1971) in his review listed thirty examples where the immobilized enzyme was reported to be more stable than the soluble form. Some of the latter improvements were due to reduced autodigestion; but in other cases there probably was improved inherent stability.

Recent reports on the stability and activity of immobilized enzymes are noted here. Neurath and Weetall (1970) reported increased activity of DNA-ase coupled to porous glass, supposedly from increases in the contact of partially degraded DNA with enzyme. Gabel *et al.* (1970) showed rather dramatically that in some cases markedly increased stability to materials such as urea, that normally cause chain-unfolding, could be obtained by immobilization; these authors obtained retention of 60% activity of trypsin in 8 M urea over seven days by coupling the enzyme to Sephadex G-200. Unbound trypsin showed no enzymatic activity in 8 M urea. Weetall (1970) has evaluated the storage stability of ficin, papain, trypsin, and glucose oxidase bound to a variety of carriers. For all of the enzymes there was at least one enzyme/support combination that exhibited 100% retention of activity after storage for up to 68 days at 4—5° C, either dry or in distilled water. He concluded that enzymes covalently bound to inorganic supports showed greater stability than those bound to organic supports. Lilly (1971) reported that immobilized β-galactosidase lost 81% of its activity during storage for three years at 2—5° C. Glassmeyer and Ogle (1971) attached trypsin to aminoethylcellulose using glutaraldehyde and reported retention of 85—90% activity when stored in water for 40 hours at room temperature or several months at 4° C. Guilbault and Das (1970) were able to immobilize cholinesterase in polyacrylamide gel such that the loss in activity was only 10% after 90 days at room temperature. When stored at 0° C between use, both cholinesterase and urease polyacrylamide gel preparations were stable and could be used repeatedly for analytical determinations with essentially no loss in activity for 80—90 days. Chang (1971a) was able to increase the stability of microencapsulated catalase by cross-linking the enzyme with glutaraldehyde after microencapsulation in collodion or nylon in the presence of high amounts of hemoglobin. At 4° C both the high concentration of hemoglobin and the cross-linking gave only 0—10% loss in activity after 45 days. At 37° C the high hemoglobin and cross-linked material retained 20% activity after nine days; while the high hemoglobin case in absence of cross-linking lost all activity after five days; and the free catalase was inactive in two days. Although there are numerous reports citing increased stability of enzymes on immobilization, there are other cases where decreased stability has been observed. It seems logical to assume that certain changes in the microenvironment around the enzyme may increase the stability of the enzyme for long-term usage. However, in general we are not able to explain *why* the stability increases or decreases upon immobilization on a particular support or better still to predict correctly the effect of a specific support on the stability of an enzyme without the need to resort to experimental testing. It seems premature to make a general statement, as a number of authors have done, that immobilization of enzymes leads to improved stability. It is still mainly a trial-and-error task, based largely on experimental testing as guided by some qualitative concepts, to obtain immobilized enzymes having improved stability. This is true especially for the inherently less stable intra-cellular enzymes.

5. Kinetics and Reactors

The anticipated increased usage of enzymes in analysis, industrial processing and medicine is based on two characteristics of enzyme catalysts, namely their specificity and their high reaction rates under relatively mild reaction conditions. In some applications the degree of specificity is the sole criterion, and the rate of reaction is of minor importance. Certain enzyme-based analytical devices and methods, as well as some of the present and contemplated uses of enzymes in medicine, fall within this category. In numerous other cases it is the combination of specificity and high reaction rates that is of interest. Other analytical devices, controlled biomedical uses of enzymes in therapy and diagnosis, and most importantly industrial processing all reside in the combined specificity/fast reaction category. It is mainly this latter category to which this section pertains. However, before delving into the performance of enzymes in a variety of reactor types and configurations, it is appropriate to review some of the parameters and methodology for the description and determination of enzyme kinetics and to discuss their application in a few examples of special biological importance. This is followed by a discussion of diffusional resistance and its significance for immobilized enzymes. The section ends with a review of numerous investigations of immobilized-enzymes in different reactor types and suggests some areas in need of additional study.

a) Enzyme Kinetics

The study of the kinetics of enzyme-catalyzed reactions has been a powerful tool for developing a better understanding of the mechanisms of action of numerous enzymes. In fact nearly all of the vast literature on enzyme kinetics has had this objective. In contrast only a small number of publications has appeared oriented towards the development of enzyme-kinetic concepts and equations for the design and optimization of enzyme reactors. The detailed reaction mechanism of a single enzyme will probably be the same whether the enzyme is functioning *in vivo* or *in vitro* assuming the same microenvironment. However, the relative concentrations and availability of enzyme and substrates in the microenvironment as well as the spatial proximity of other relevant enzymes, interfaces or inhibitory materials may be quite different for the *in vitro* and *in vivo* cases and even different for the *in vitro* kinetic mechanistic studies of the biochemist and the enzyme-reactor studies of the engineer. The subject of enzyme kinetics has been reviewed from the viewpoint of the biochemist by Cleland (1967) and that of the engineer by Carbonnel and Kostin (1972).

The most widely adopted model for enzyme kinetics is the one-enzyme one-substrate, steady-state approach of Briggs and Haldane (1925) described earlier in the introduction. The assumption that the concentration of the enzyme substrate complex reaches a steady-state value of course is an approximation since the substrate level, and thus the concentration as well as the rate of change of concentration of enzyme substrate complex, changes as the reaction proceeds. Heineken, Tsuchiya, and Aris (1967) have shown from the theory of single perturbations that the main assumption needed for the steady-state approach is that the ratio of total initial enzyme to substrate concentrations must be very small. This assumption often is ignored when Eq. (1) is modified to remove the originally imposed assumption that $[S] \approx [S]_0$ to give Eq. (3) and the latter integrated over a sufficient time interval such that the ratio of $[E]_0$ to $[S]$ does not remain very small.

$$\frac{d[S]}{dt} = \frac{V_m[S]}{K_m + [S]} \tag{3}$$

Let: $[S] = [S]_0 - [X]$

where $[X]$ is the concentration of substrate converted to product. The integrated equation becomes Eq. (4):

$$\ln\left[\frac{[S]_0}{[S]_0 - [X]}\right] = \frac{V_m t}{K_m} - \frac{[X]}{K_m} \tag{4}$$

In the Briggs-Haldane approach (commonly called the Michaelis-Menten equation) the parameters V_m and K_m must be evaluated by fitting Eq. (1) or (4) to experimental data. Schwert (1969) has discussed the estimation of these two parameters using the integrated Michaelis-Menten equation [Eq. (4)]; while the more traditional methods of linearizing the hyperbolic Michaelis-Menten equation [Eq. (1)] to plot $1/v_0$ versus $1/[S]_0$, $[S]_0/v_0$ versus $[S]_0$, or v_0 versus $v_0/[S]_0$ are described in most texts on enzyme kinetics. Each of these three methods for linearizing Eq. (1) is equally valid so long as experimental data are available to cover a wide range of initial substrate concentrations and velocities. Many authors neglect to refer to the reaction velocity and substrate concentration as initial values in the Michaelis-Menten equation; again such an omission is valid if the ratio of $[E]_0$ to $[S]$ is very small. Cha (1970) has shown by calculation that when the concentrations of $[E]_0$ and $[S]_0$ both are within 10-fold of K_m, the relative error in the Michaelis-Menten equation (as compared to the true rate expression) ranged from 0.08% for $[S]_0 = 10\,K_m$ and $[E]_0 = 0.1\,K_m$ to 900% for $[S]_0 = 0.1\,K_m$ and $[E]_0 = 10\,K_m$. Lee and Wilson (1971) have suggested that the error in estimating K_m and V_{max} with up to 50% substrate utilization could be reduced to less than 4% by rearranging the integrated form of the Michaelis-Menten equation [Eq. (4)] to Eq. (5) and modifying to Eq. (6).

$$\frac{1}{[X]/t} = \frac{1}{V_m} + \frac{K_m}{V_m[X]} \ln \frac{[S]_0}{[S]_0 - [X]} \tag{5}$$

$$\frac{1}{\bar{v}} = \frac{1}{V_m} + \frac{K_m}{V_m} \frac{1}{[\bar{S}]} \tag{6}$$

where: $\bar{v} = \dfrac{[X]}{t}$

$$[\bar{S}] = [S]_0 - \frac{[X]}{2} \approx \frac{1}{[X]} \ln \frac{[S]_0}{[S]_0 - [X]} .$$

Eq. (6) is a modified Lineweaver-Burk linearized form of the Michaelis-Menten equation, with a plot of $1/\bar{v}$ vs. $1/[\bar{S}]$ used to obtain K_m and V_m from experimental data. Lee and Wilson (1971) cited the Theorem of the Mean to show that a value of $[S]$ existed in the interval between $[S]_0$ and $\{[S]_0 - [X]\}$ where $d[S]/dt$ equaled $[X]/t$ and that this value of $[S]$ was approximated closely by the arithmetic mean of $[S]_0$ and $\{[S]_0 - [X]\}$ or by $[S]_0 - [X]/2$.

Since often it is difficult to extrapolate a plot of reaction extent *versus* time so as to get the initial reaction rate (slope at time zero), Johnston and Diven (1969) have suggested a method for obtaining initial rates using an integrated rate equation. Their method is based on the reversible reaction $ES \rightleftharpoons E + P$ and assumes $[S]_0 \gg [E]_0$, a steady-state for $[ES]$ and for $[E]$, and a knowledge of the concentration of product or substrate at equilibrium.

The Michaelis constant was defined in Section 1b as:

$$K_m = \frac{k_2 + k_3}{k_1} .$$

In certain enzyme-reactor design studies as well as mechanistic investigations, it is necessary to obtain values for the individual rate constants k_2, k_3, and k_1. These values cannot be obtained from steady-state data (in terms of the formation of the enzyme/substrate complex) but must be developed from pre-steady-state measurements (a time span usually of a few milliseconds). Darvey (1968) has provided a solution for the transient phase of a general enzyme mechanism; the general techniques for obtaining the values for the individual rate constants are described in most advanced texts on enzyme kinetics. Fast reaction methods such as stopped-flow and relaxation techniques are employed for pre-steady-state measurements. In the latter method, an equilibrium enzyme/substrate/product system is rapidly perturbed and the concentration of one of the species or else some other reaction-dependent variable is followed during the return to equilibrium. The technique is of use mainly for detailed mechanistic studies. The stopped-flow and relaxation methods have been described by Gibson (1966), Eigen and DeMaeyer (1963), and Chock (1971) among others.

An alternate procedure for obtaining values for the rate constants and indirectly for the Michaelis constant or the maximum velocity, is by numerical solution of the differential equations for the rate of change of substrate, enzyme, enzyme/substrate complex, and product. Tanner (1972) has suggested using Picard's iteration technique followed by data-fitting with least-squares orthogonal polynomials; a comparison of the Picard and orthogonal polynomial coefficients provides an estimate of the kinetic parameters of interest. Not all enzymes have kinetics that follow the hyperbolic equation of Michaelis and Menten; allosteric enzymes for instance may show a sigmoidal relationship for velocity versus substrate concentration. Weiker et al. (1970) have described a computer program for estimating kinetic constants for sigmoidal systems. The method of Tanner (1972) is applicable to non-linear systems in general. Darvey (1969) has suggested a method for obtaining the form of the kinetic rate expression for non-linear systems; his method depends on obtaining initial steady-state velocities at very low and very high reactant concentrations.

In both reactor design studies and mechanistic investigations the possibility of competition of several substrates for a given enzyme as well as the presence of multi-enzyme systems in series (or parallel) greatly increases the difficulty of analysis of the kinetic data. Pocklington and Jeffery (1969) have provided an example of 5-α-androstan-3-one and 5-α-androstan-3, 16-dione competing for a hydroxysteroid-NAD oxidoreductase for the case where only binary enzyme/substrate complexes are formed (no ternary complex). They showed that the total initial rate of reaction may be greater than or lie between the rates for either substrate alone depending on the concentration of the substrate with the higher maximum velocity. Methods for analysis of the kinetics of multi-enzyme systems have been discussed by Savageau (1969a, b) who suggested using a power-law approximation for non-linear portions; Rhoads et al. (1968a), who replaced the differential rate equations with an algebraic rate law; Rhoads and Pring (1968), who developed two computer programs for derivation of the rate equations utilizing the steady-state approximation for all of the enzyme species; and Vergonet and Berendsen (1970), who also utilized the steady-state approximation. The development of simulation techniques and methods for estimation of kinetic parameters as well as methods for the solution of kinetic rate expressions for multi-enzyme and/or multi-substrate systems may provide more powerful tools for mechanistic studies of enzyme systems and for the design and optimization of multi-enzyme reactors. However, a limiting factor which many times appears to be overlooked is our general inability to obtain suitable experimental data for definitive testing of various assumptions and kinetic models of specific enzyme systems. This is true especially for *in vivo* studies and likely will be an important limitation of efforts to optimize the design and performance of multienzyme flow reactors.

An interesting example of the type of phenomena that may be found with multi-enzyme systems is the continuous or damped oscillation of reactant, product, or enzyme concentrations. Hess and Boiteux (1971) have reviewed the experimental evidence for such oscillations and have cited a number of attempts to develop mathematical models to explain the phenomenon in specific cases. Such model development obviously will be of use in explaining the dynamics of multi-enzyme systems *in vivo* (provided suitable experimental data can be obtained) and especially in the analysis of multi-enzyme flow reactors *in vitro*. Another phenomenon is the evidence of hysteresis in sequential enzymatic reactions when a product rate of formation is plotted as a function of a precursor concentration at least one reaction step removed from the product-forming step (Tanner, 1971).

The traditional approach to developing quantitative relationships for the time-course of velocities and concentrations for enzyme reactions has been deterministic in nature, in that the equations (such as the Michaelis-Menten type) have utilized continuous time-dependent concentrations as variables. Staff (1970) and Smith (1971) recently have extended the stochastic approach to the development of the Michaelis-Menten equation. They also applied this approach to an open linear system with an enzyme reaction where the concentrations of substrate and of a reversible competitive inhibitor were maintained constant and product removal was by a first-order process. The latter work may be of special importance under conditions where enzyme reactions occur in small systems with only a small number of molecules present such that random fluctuations in reaction variables take on practical significance.

It was mentioned in the previous section that immobilization often results in a modification to kinetic parameters such as K_m and V_m. In the absence of significant diffusional resistances, the above kinetic parameters may be modified upon enzyme immobilization by changes in the pH, electrical potential, or relative concentrations in the microenvironment or by conformational and steric restraints set up by the enzyme/

support geometry. In their review, Goldman, Goldstein, and Katchalski (1971) cited a number of specific examples of changes in kinetic parameters due to enzyme immobilization. A change in the activity *vs*. pH response perhaps was typical. However, a number of cases of unchanged kinetic parameters also have been demonstrated for immobilized enzymes. Some examples are unchanged K_m but changed V_m for ribonuclease T_i bound to Sepharose by Lee (1971), trapped phosphoglycerate mutase by Bernfeld *et al*. (1969), trapped glucose oxidase or lactic dehydrogenase by Hicks and Updike (1966), and bound glucose oxidase by Weetall and Hersh (1970). In general the effect on kinetic parameters of immobilization of a specific enzyme by a certain technique cannot be predicted at present. For mechanistic studies, this inability to predict the effects of immobilization on kinetic parameters is a major limitation to the development of more highly active immobilized-enzyme systems, especially for the enzymes of intracellular origin. However, in the design and analysis of immobilized-enzyme reactors diffusional resistances often mask changes in diffusion-free kinetic parameters and can play a more significant role in attaining high reactor performance.

b) Diffusional Resistances

The wide variety of methods and media for the immobilization of enzymes provides several special situations where the rates of transport of substrate or product to or from the enzyme molecule may significantly affect the overall obtainable rate of chemical transformation. The main cases, pictured in Fig. 4, are as follows:

1. Diffusion through external "stagnant" films adjacent to enzyme support materials.
2. Diffusion through polymer and gel support media.
3. Diffusion in and out of narrow internal enzyme-containing pores in porous supports.
4. Passive diffusion and facilitated transport through enzyme-containing membranes.
5. Interphase transport for enzymes immobilized at an interface.

In a number of studies with immobilized enzymes the rates of diffusion of reaction species have had a major and sometimes limiting effect on the overall rate of conversion of substrate to product. Yet very few studies have been carried out to quantitatively define the influence of diffusional resistances on immobilized enzyme kinetics. Ford *et al*. (1972) have described an experimental "differential-type" reactor system for measuring the kinetic parameters of immobilized enzyme reactions in the absence of external diffusional resistances. Hornby, Lilly, and Crook (1968) observed an increase of 23—1000% in K_m for covalent coupling of ATP-

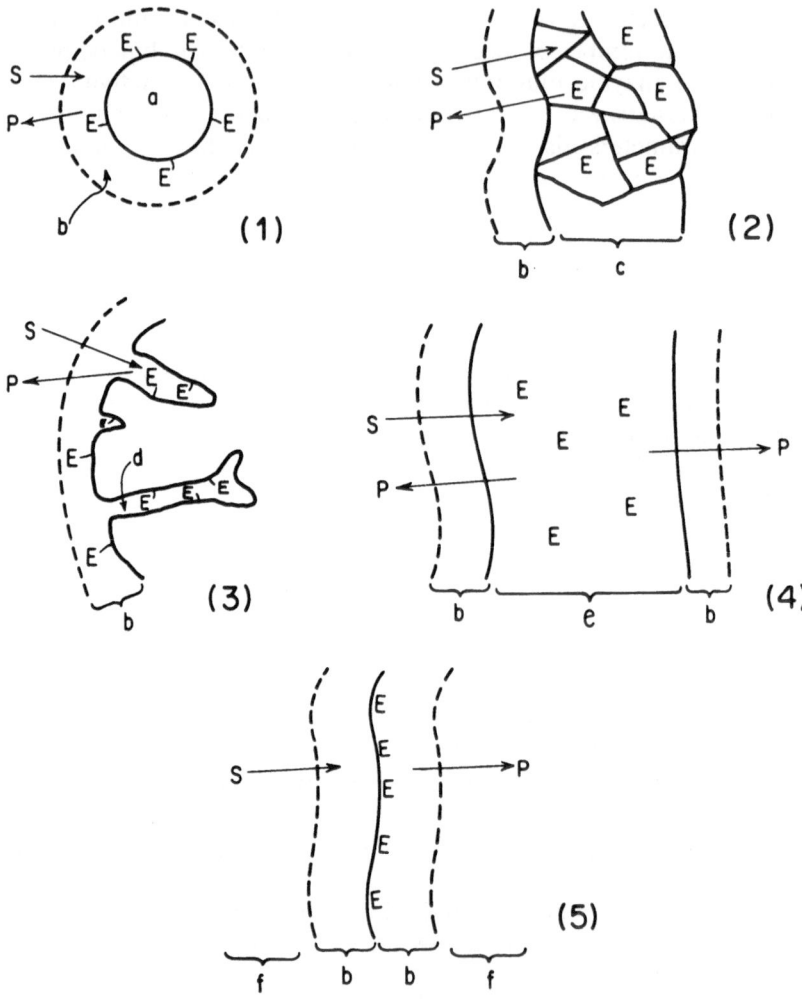

Fig. 4. Sources of diffusional resistances. a solid particle interior. b stagnant film. c gel or support media. d pores. e membrane interior. f bulk of fluid. E enzyme, P product, S substrate. 1 diffusion through external film. 2 diffusion through polymer and gel support. 3 diffusion into pores. 4 passive diffusion and facilitated transport. 5 interphase transport, E at interface

creatine phosphotransferase on charged cellulosic supports. They developed Eq.(7) for the apparent Michaelis constant (K'_m) by equating the steady-state rate of reaction (Michaelis-Menten equation) to the rate of transport of substrate. They defined the latter as the sum of transport due to a gradient of electrical potential plus transport due to a gradient of substrate concentration across an external stagnant layer:

$$K'_m = \left(K_m + \frac{yV_m}{D}\right)\left(\frac{RT}{RT - {}_3YF\,\mathrm{grad}\psi}\right) \qquad (7)$$

where K'_m = apparent Michaelis constant, K_m = "true" Michaelis constant, y = thickness of stagnant diffusion layer, D = diffusion coefficient, R = gas constant, T = absolute temperature, ${}_3$ = charge, ψ = electrical potential, and F = Faraday constant. The authors found qualitative agreement of their data with Eq. (7). In another study Boguslaski and Janik (1971) encapsulated carbonic anhydrase in thin-walled (ca. 200 Å) microcapsules of 70—110 μm diameter and found K'_m to be 3—4 times greater than for the enzyme in solution; although some of this difference may have been due to incomplete encapsulation of the enzyme. Dye molecules with a molecular weight of 1000 diffused rapidly through the walls of the microcapsules. In most of the above examples, diffusional resistances were caused mainly by external stagnant layers, or by the support matrix.

The quantitative description of diffusional resistances by stagnant films around enzyme-containing particles should be subject to the same approach found in the chemical engineering literature for diffusion through stagnant films, although in numerous cases the overall rate may depend both on the rates of diffusion and reaction. Schurr (1970) has presented a theoretical analysis of diffusion as it effects enzymes in solution or adsorbed on the surface of large particles. He concluded that the activity of enzymes in solution generally was independent of solution viscosity (and thus of diffusion rates); for adsorbed enzymes the rates either decreased or remained constant unless the enzyme became saturated with substrate at which point the rate became independent of diffusional limitations. With enzymes trapped in gels or tightly cross-linked polymer matrices, resistance to diffusion within the gel or polymer matrix frequently becomes the rate-controlling step, especially for substrates of larger molecular weight. In assaying gel- or matrix-trapped enzyme systems for activity it is often necessary to make measurements at a series of decreasing particle sizes for determining the percentage of diffusion-free enzyme activity retained in the immobilized system. Laurent (1971) has developed a theoretical treatment for enzyme kinetics in polymer media and applied it to three experimental systems. His method makes use of activity coefficients in the Michaelis-Menten equation; he defined the coefficients as the ratio of K'_m (apparent constant) to K_m (true constant). The following systems gave reasonable values for the activity coefficients: a) degradation of hyaluronic acid by hyaluronate lyase in polyethylene glycol solution, b) the lactate dehydrogenase reaction in dextran solution, and c) the cleavage of an ester by trypsin inhibited by serum albumin in dextran solution. Ceska (1971) studied the effect of interfacial area on the hydrolysis of water-insoluble starch by α-amylase.

The use of porous particles such as glass beads provides a convenient method for increasing the surface area for enzyme attachment. However the internal pore surfaces become relatively inaccessible to enzyme or substrate as the pore diameter is decreased or the size of the molecules is increased. Messing (1970) has shown that less papain per gram of glass is bound with pore diameters 50,000 Å and 92 Å as compared to 900 Å. Weibel and Bright (1971) have studied the kinetics of glucose oxidase covalently bound to porous glass beads having 400—500 Å diameter pores; they found only 6% of the immobilized enzyme to be catalytically active under test conditions and suggested that the remaining 94% was immobilized inside the pores and made inaccessible to substrate by pore-diffusion limitations. In a later study Weibel and Humphrey (1971) suggested that 90—95% of the internal pore surface area was not utilized since calculations showed 90—95% of the initial oxygen concentration at the pore entrance was depleted within a distance of only 7,500 Å into the pores. Thus within the pore the $FADH_2$ moiety of the enzyme was prevented from being reoxidized. The authors suggested further that under this type of diffusional limitation to utilize all of the enzyme would require a first-order rate constant of 1—10 sec^{-1} or a second-order constant of 10^4 molar^{-1} sec^{-1}. Very little has been published on methods for predicting the extent of pore-diffusional limitations for immobilized enzyme catalysts. Ford et al. (1972) have presented expressions without derivation for a parameter called the Thiele modulus for the limiting cases of $K_m \ll [S]$ and $K_m \gg [S]$. These authors also cited a comparison paper in press on diffusional aspects of immobilized enzyme kinetics. Ollis (1971) presented a theoretical analysis of the effects of pore-diffusion control on the apparent thermal stability of reversibly or irreversibly denaturable enzymes when $[S] \ll K_m$.

The general concept of the Thiele modulus and the "effectiveness factor" is introduced here for the case of Michaelis-Menten kinetics to show the predictive capability of the method. The treatment here is for porous spherical beads with enzyme attached mainly to the surfaces of the internal pores. For the spherical particle shown in Fig. 5, a material balance is written equating the net steady-state diffusion of substrate into the differential element of volume of thickness dr and radius r to the substrate reacted within the volume element.

Rate of substrate diffusion IN =

$$-D_e \frac{d[S]}{dr} (4\pi r^2).$$

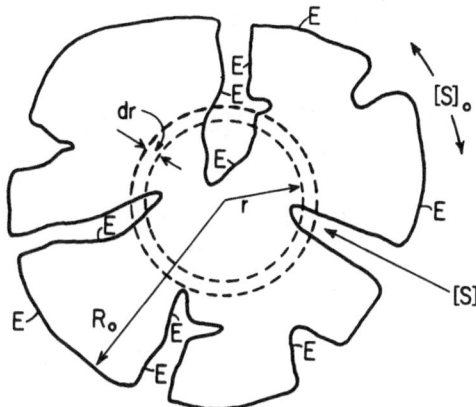

Fig. 5. Pore diffusion in spherical pellett. $[S]$ substrate concentration within pore. $[S]_0$ substrate concentration at pore mouth. r radial distance. R_0 radius of sphere. Neglect stagnant layer since assume pore diffusion rate-controlling

Rate of substrate diffusion OUT (towards center of particle) =

$$-D_e \left[4\pi r^2 \frac{d[S]}{dr} + \frac{d}{dr} \left(4\pi r^2 \frac{d[S]}{dr} \right) dr \right].$$

Rate of substrate reaction =

$$-4\pi r^2 (dr) \varrho_p A_g \left(\frac{V_m[S]}{K_m + [S]} \right)$$

where: ϱ_p = density of particle (solid and voids),
       A_g = surface area per weight of catalyst.

The net equation becomes:

$$\frac{d^2[S]}{dr^2} + \frac{2}{r} \frac{d[S]}{dr} = \frac{\varrho_p A_g}{D_e} \left(\frac{V_m[S]}{K_m + [S]} \right). \tag{8}$$

By redefining $[S]$ and r in terms of dimensionless variables, the resulting parameters in general become more useful.
Let:

$$[S]^* = \frac{[S]}{[S]_0}, \quad \bar{r} = \frac{r}{R_0}.$$

Thus, Eq. (8) becomes upon substitution:

$$\frac{d^2[S]^*}{d\bar{r}^2} + \frac{2}{\bar{r}} \frac{d[S]^*}{d\bar{r}} = \frac{\varrho_p A_g R_0^2}{D_e} \left(\frac{V_m[S]^*}{K_m + [S]_0[S]^*} \right). \tag{9}$$

For the two limiting cases, the right side of Eq. (9) reduces to:

1) $K_m \gg [S]_0 [S]^*$

$$\left(\frac{\varrho_p A_g R_0^2 V_m}{D_e K_m} \right) [S]^* = h_1^2 [S]^*$$

where:

$$h_1 = R_0 \left(\frac{\varrho_p A_g V_m}{D_e K_m} \right)^{1/2}. \tag{10}$$

2) $K_m \ll [S]_0 [S]^*$

$$\left(\frac{\varrho_p A_g R_0^2 V_m}{D_e [S]_0} \right) = h_0^2$$

where:

$$h_0 = R_0 \left(\frac{\varrho_p A_g V_m}{D_e [S]_0} \right)^{1/2}. \tag{11}$$

By definition h_1 and h_0 are the Thiele moduli for the indicated conditions.

The Effectiveness Factor, $\mathfrak{E}$, is defined as follows:

$$\mathfrak{E} = \frac{\text{Overall reaction rate with pore-diffusion limitations}}{\text{Overall reaction rate without diffusion limitations}}.$$

Thus $\mathfrak{E}$ is the fraction of total surface (and thus of enzyme assuming a uniform distribution of enzyme over the surface) available for use by the substrate at the concentration of reactant at the pore mouth. Following the extensive discussion of this method by Wheeler (1951), values of $\mathfrak{E}_1$ can be calculated using Eq. (9) with the assumption of $K_m \gg [S]_0 [S]^*$ and the following relationship:

$$\mathfrak{E}_1 = \frac{-4\pi R_0^2 D_e \left(\dfrac{d[S]}{dr} \right) \text{at } r = R_0}{\dfrac{4}{3} \pi R_0^3 \varrho_p A_g \dfrac{V_m [S]_0}{K_m}} \tag{12}$$

or in dimensionless variables

$$\mathfrak{E}_1 = \frac{-3 \left(\dfrac{d[S]^*}{d\bar{r}} \right) \text{at } \bar{r} = 1}{h_1^2}. \tag{13}$$

Thus, from a knowledge of the physical aspects of the support, the enzyme kinetic parameters of V_m and K_m, the effective diffusion coefficient for the substrate within the pores, and the substrate concentration gradient at the pore mouth one can predict the fraction of total surface (and thus enzyme) available to the substrate. Additional predictive equations can be developed using other expressions instead of the Michaelis-Menten form for the reaction term. In cases of strong pore-diffusional limitations, it is doubtful if the Michaelis-Menten equation can be used without considerable error since the requirement that $[E]_o \ll [S]$ would not be satisfied throughout the entire region of the reaction.

c) Reactor Studies

With the ease of enzyme recovery and reuse brought by immobilization, there has been considerable interest in studying the performance of immobilized enzymes in a wide variety of reactor types. Stirred tanks, packed beds, hollow tubes, fuel-cell and other electrode reactors, membrane units, and other more novel devices have been utilized with sporadic attempts to define the more significant reactor variables. Lilly, Hornby, and Crook (1966) studied the effects of flow-rate and initial substrate concentration on the hydrolysis of an ester by immobilized ficin in a packed bed. Using the integrated Michaelis equation and assuming plug flow, the authors got good experimental agreement for the fraction of substrate reacted as described by Eq. (14).

$$f[S]_0 = K'_m \ln(1 - f) + \frac{V_m B}{Q} \tag{14}$$

where: f = fraction of substrate reacted
K'_m = apparent K_m,
Q = flow-rate,
B = void fraction.

In a later study Lilly and Sharp (1968) developed a similar relationship [Eq. (15)] for immobilized enzymes in a continuous-flow stirred-tank reactor.

$$f[S]_0 = -K'_m \left(\frac{f}{1-f}\right) + \frac{V_m B}{Q}. \tag{15}$$

K_m was affected by flow-rate in both reactors while V_m was flow-dependent only in the stirred-tank case. The problem of getting higher flow-rates than could conveniently be obtained with packed column operation was approached by attaching enzymes to porous sheets of filter paper with flow of substrate through the sheets. Wilson, Kay, and Lilly (1968) developed such a reactor system with pyruvate kinase in series with lactate dehydrogenase. The extent of the conversion of pyruvate to lactate depended on the order of the enzyme sheets as well as the flow-rate; the values for K'_m and V'_m for the kinase sheet were relatively constant with flow, indicating a rather constant diffusional resistance. Kay *et al.* (1968) have summarized the effect of flow-rate on conversion for several different enzymes attached to flow-through sheets. With β-galactosidase attached to porous cellulose sheets Sharp, Kay, and Lilly (1969) found a high dependency of K'_m and V'_m on flow-rate for some of the substrates.

Weetall and Baum (1970) found the conversion of L-aminoacids to be independent of flow-rate for L-aminoacid oxidase immobilized on porous glass beads. In an extensive study of the use of glucoamylase immobilized on glass beads for converting corn starch to dextrose, Weetall and Havewala (1972) used both a packed-bed and a stirred-tank reactor. They examined the effects of flow-rate, pore size, immobilization technique, etc. and even with incorporation of a term for enzyme inactivation concluded that the simple Michaelis-Menten model was not sufficient to explain their experimental system.

O'Neill, Dunnill, and Lilly (1971) used amyloglucosidase bound to cellulosic supports to convert maltose to glucose. Flow-dependent kinetic parameters were observed both with packed-bed and stirred-tank operation. The packed-bed system appeared to follow first-order rather than Michaelis-Menten kinetics; and at high flow-rates the diffusional effects were minimal. In a related study, Smiley (1971) operated a continuous 4-liter stirred-tank reactor to convert liquified starch to glucose with glucoamylase complexed to DEAE-cellulose. Substrate concentrations up to 30% were reacted to produce glucose at rates exceeding 25 mg $\text{min}^{-1} \cdot \text{l}^{-1}$ for 3—4 weeks. No evidence of enzyme washout from the reactor was observed.

Butterworth *et al.* (1970) have studied an ultrafiltration enzyme reactor and Porter (1972) has described a thin-channel ultrafiltration system for retention and recycle of enzyme catalysts. Bowski *et al.* (1972) used the latter type-reactor to study the continuous hydrolysis of sucrose by invertase and developed a mathematical model based on the kinetic model of Bowski *et al.* (1971), that agreed well with experimental data. Another example using an ultrafiltration reactor with recycle of enzyme was by Ghose and Kostick (1970); they hydrolyzed a suspension of cellulose to glucose with cellulase and developed an experimental model system for further study of this two-phase reaction. Hornby and Filippusson (1970) attached trypsin to the inside of nylon Type 6 tubes and passed substrate through the tubes; they found a strong flow-dependence for the hydrolysis of an ester. Rony (1971) described a hollow-tube reactor in which soluble or immobilized enzyme could be placed in the tubes and the latter sealed at both ends; substrate and product could diffuse through the tube walls. Rony presented a mathematical analysis of diffusional control for sealed-tube enzyme reactors and compared the theoretical expressions with similar ones for packed-bed and stirred-tank enzyme reactors. His analysis in part was based on defining an overall effectiveness factor such that its reciprocal, the total resistance to mass transfer, could be equated to the sum of the reciprocals of the effectiveness factors for the individual resistances of the internal and external liquid films and the membrane.

Instead of immobilizing crude or purified enzymes, Mosbach and Larsson (1970) trapped *Curvularia lunata* in crosslinked polyacrylamide gel and used the resulting system to convert steroid Reichstein compound S *via* 11-β-hydroxylation to cortisol (followed by conversion of cortisol to prednisolone) in a continuous-flow column using gel-immobilized steroid- Δ^1-dehydrogenase. In a related study, Franks (1972) trapped *Streptococcus faecalis* in crosslinked acrylamide gel and used it in a column to convert arginine to ornithine (*via* arginine deaminase, ornithine transcarbamylase, and carbamyl kinase).

In another variation O'Neill, Wykes, Dunnill, and Lilly (1971) attached chymotrypsin to dextran giving a soluble system and found increased activity over the same enzyme attached to insoluble DEAE-cellulose. The soluble bound enzyme was operated continuously in an ultrafiltration cell for two weeks at a feed rate of 2.50 ml/day of 0.5% caesin; no loss in enzyme activity was observed.

Some enzymes are inhibited by high substrate concentrations. O'Neill (1971) and O'Neill, Lilly, and Rowe (1971) have shown theoretically that single-substrate systems that undergo substrate inhibition may have, in the presence of mass-transfer effects, more than one set of conditions where $d[S]/dt$ is zero (steady-state) for a continuous-flow stirred-tank reactor. The regions of unstable steady-state conditions occur mainly at low ratios of $k/[S]_0$ and in the range of 50—100% conversion. Kobayaski and Moo-Young (1971) developed a theoretical treatment for combined Michaelis-Menten reaction and diffusion through a stagnant film and obtained approximate solutions for the limiting cases of perfect mixing and plug flow.

In the area of electrochemical enzyme reactors Wingard and Liu (1969) have presented a theoretical calculation of the expected steady-state current for a glucose-oxidase-catalyzed fuel-cell. The use of constant current voltametry for the study of enzyme-electrode parameters has been described by Wingard, Liu, and Nagda (1971) and applied to glucose oxidase in polyacrylamide gel.

An area of special interest for future applications of immobilized enzymes is the use of multi-enzyme systems to carry out a sequence of transformations. One such reactor system by Wilson, Kay, and Lilly (1968) used pyruvate kinase in series with lactate dehydrogenase, each enzyme attached to a separate flow-through sheet. Mosbach and Mattiasson (1970) used a different approach and coupled two and later three

enzymes (Mattiasson and Mosbach, 1971) to the same support. Hexokinase and glucose-6-phosphate dehydrogenase were covalently bound
simultaneously to a solid polymeric support. The formation of NADPH
plus H^+ was monitored after addition of excess glucose, ATP, and
$NADP^+$. For the immobilized two-enzyme system, the rate of formation
of NADPH plus H^+ was increased by 40—140% over that from the two
enzymes in solution. Presumably the concentration of glucose-6-phos-
phate, formed by hexokinase and glucose, was higher in the microenvironment of the bound enzyme than for the unbound enzyme (due possibly to unstirred layers). Assuming that the second enzyme (the dehydrogenase) was close to the product of the first reaction, the second reaction
would proceed at a rate higher than with the unbound enzymes. In the
three-enzyme system β-galactosidase was added to the above and lactose, ATP, and $NADP^+$ used as substrates. The third enzyme converted
lactose to glucose and galactose, thus providing the reactant for the rest
of the sequence. Again the overall rate was higher for the bound than
soluble system. Goldman and Katchalski (1971) presented a theoretical
analysis for two enzymes bound to the surface of an impermeable support. The reaction sequence was as follows:

$$S \xrightarrow[E_1]{} P_1 \xrightarrow[E_2]{} P_2 \,.$$

The authors assumed first-order kinetics with no products present initially in the bulk of the solution. They concluded that the presence of
unstirred layers kept product P_1 in the vicinity of the support surface so
that reaction 2 proceeded more rapidly than if both enzymes were free in
solution. However, the relative amounts of each enzyme needed to be
considered (as Mattiasson and Mosbach (1971) also pointed out) since
the ratio of E_1/E_2 could vary the overall reaction rate over wide limits.
An increasing number of multienzyme complexes have been found in
living cells for the catalysis of a sequence of perhaps two to six reactions.
Several systems described by Reed and Cox (1970) appear to consist of
particles containing definite amounts of each enzyme but free of lipids,
nucleic acids, or membrane components. Another more complex class of
in vivo multienzyme systems comprises those associated with energy
transfer, synthesis, and metabolic processes carried out in subcellular
organelles. These latter multienzyme systems in many cases include
membrane-bound enzymes that require phospholipids or other types of
lipoprotein complexes for catalytic activity. Although our understanding
of the functioning and rationale of membrane-bound enzymes in cellular
processes is rather meager, we may find possibilities for the development
of useful *in vitro* enzymatic reaction systems through the study of both *in
vivo* and *in vitro* membrane-bound enzymes.

d) Membrane/Enzyme Combinations

The literature on the structure and function of membrane-bound or membrane-trapped enzymes is overwhelming. However, much of this literature pertains to the mechanisms of action of membrane enzymatic processes in subcellular particles; and even though engineering methodology may be helpful here, the required detailed knowledge is so large that very few engineering-oriented investigators will accept the callenge to work in this area. Rather, the study of the combined reaction/transport kinetics of enzymes associated with synthetic membranes appears more appropriate for the engineering-oriented investigator. Although this latter topic has been examined both theoretically and experimentally by a number of investigators, the understanding as well as the development of useful enzyme/membrane processes and devices is only beginning.

McLaren and Packer (1970) have presented a pertinent review of enzyme reactions in heterogeneous systems, such as solid (E)/solution (S) (carboxypeptidase), gel (E)/solution (S) (cross-linked trypsin), surface (E)/surface (S) (ATP-ase), solution (E)/solid (S) (chitinase), or solution (E)/liquid (S) (lipase). Such systems involving enzyme activity at interfaces are pertinent to the study of certain types of membrane/enzyme configurations. For example Brown, Selegny, Mink, and Thomas (1970) have prepared films of carbonic anhydrase glutaraldehyde on silastic and observed increased (facilitated) transport of carbon dioxide across the membrane. The enzyme exerted its greatest influence at the solution membrane interface since the change in the concentration profile across the membrane in the presence and absence of enzyme was greatest at the interfaces.

Goldman, Goldstein, and Katchalski (1971) and Katchalski, Silman, and Goldman (1971) have presented two similar reviews on the theoretical analysis of the kinetics of an enzyme distributed homogeneously throughout the matrix of a substrate-and product-permeable artificial membrane. They have included first-order as well as Michaelis-Menten kinetics, the presence of stagnant films of fluid on each side of the membrane, and a variation of pH within the membrane due to the enzymatic reaction being studied. These reviews are based on the original theoretical kinetic analysis by Goldman, Kedem, and Katchalski (1968) and the latter experimental study by the same authors (1971). In the last-mentioned paper, the authors reported that the diffusional resistance of the stagnant layers could be neglected in certain cases, but not with high enzyme activity and low K'_m. Under certain conditions both K'_m and V'_m for reaction within the enzyme membrane system were influenced by membrane thickness, extent of stagnant layers, and other factors.

Numerous papers have appeared recently in the chemical engineering and biophysics literature on facilitated membrane transport caused by reversible association of reactant with a carrier within the membrane or by reaction within the membrane. But only a few examples of transport facilitated by enzyme-containing membranes have been reported. Selegny, Broun, and Thomas (1971) have prepared reticulated (i.e. resembling a network) and coreticulated enzyme membranes by crosslinking an enzyme with a difunctional reagent such as glutaraldehyde either directly (reticulated) or along with an inactive protein like albumin (coreticulated). The resulting matrix was essentially homogeneous for the spatial distribution of enzyme. The authors listed 18 enzymes they have immobilized by this method and studied. Under steady-state conditions the rate of diffusion at any point in the membrane is equal to the rate of reaction as shown by Eq. (16).

$$D_e \frac{d^2[S]}{dW^2} = -k_3 \frac{[E]_0[S]}{K_m + [S]}. \tag{16}$$

For the case where the membrane was placed between two different substrate solutions, Eq. (16) could be integrated to give Eq. (17). The latter presented a useful method for the direct experimental determination of K_m within the membrane.

$$(J_1)^2 - (J_2)^2 = 2k_3[E]_0 D_e \left[[S]_1 - [S]_2 - K_m \log \frac{[S]_1 + K_m}{[S]_2 + K_m} \right] \tag{17}$$

where J_1 is the flux of substrate entering the membrane and J_2 leaving and $[S]_1$ and $[S]_2$ are the substrate concentrations in the bulk of the fluids on the two sides of the membrane. With the same substrate concentration on both sides of the membrane an approximate solution to Eq. (16) could be obtained through a McLaurin series and the definition of several dimensionless parameters, α and σ

$$\alpha = \frac{[S]}{K_m} \quad \text{and} \quad \sigma = \frac{k_3[E]_0 e^2}{K_m D_e}$$

where e is the membrane thickness. When the enzyme was in solution or attached to the membrane surface ($e = 0$, $\sigma = 0$), the activity/substrate concentration curves followed the hyperbolic Michaelis-Menten form. However, as σ increased the activity/substrate concentration relationship became increasingly sigmoidal in form. The authors cited several experimental studies that showed agreement with these theoretical expressions. In the case of the non-steady state, Eq. (16) became [Eq. (18)]:

$$\frac{\partial[S]}{\partial t} = D_e \frac{\partial^2[S]}{\partial x^2} - k_3 \frac{[E]_0[S]}{K_m + [S]} \tag{18}$$

which is a non-linear second-order equation without a general analytical solution. Selegny *et al.* (1971) obtained numerical solutions for Eq. (18) and the corresponding expression for the product $[P]$ concentration, using the conditions that initially the membrane was void of substrate or product and that for $t > 0$, $[S] = [S]_1$ and $[P] = [P]_1$ at membrane "face 1" and $[S] = [S]_2$ and $[P] = [P]_2$ at membrane "face 2". The numerical solutions reduced to those previously obtained for the steady-state case.

In addition to the study by Broun *et al.* (1970), Otto and Quinn (1971) have examined the facilitated transport of carbon dioxide through thin membrane layers of bicarbonate solution with metal-ion catalysis and both in the presence and absence of carbonic anhydrase. At low enzyme concentrations the approach to equilibrium increased linearly with added enzyme, as expected. However, at carbonic anhydrase levels greater than $0.5 \text{ mg} \cdot \text{ml}^{-1}$ the fractional approach to equilibrium decreased to some extent and then held constant with added enzyme. Supporting data from stopped-flow experiments by Otto and Quinn (1972) led to the conclusion that inhibition by chloride and bicarbonate anions caused the non-linearity between enzyme concentration and activity.

Vieth and co-workers (1972) have described a novel flow-reactor, consisting of a collagen membrane impregnated with urease or lysozyme. The membrane was rolled into a spiral with spaces between consecutive turns; substrate solution was flowed along the axis of the spiral through the opening between turns. The unit retained 25% of its original activity over a period of two weeks. In another study, Julliard *et al.* (1971) bound glutamate dehydrogenase to a matrix of collagen; double reciprocal plots with NADH plus H^+ as substrate were non-linear for the bound enzyme but linear for the same system in solution.

Li (1971) has developed a novel separation technique based on liquid membranes composed of an immiscible solvent (with water) and surfactants. Shrier (1972) has described the use of such liquid membranes to encapsulate enzymes as an aid in enzyme recovery and re-use and in providing a large interfacial area for substrate transfer to the encapsulated enzyme.

e) Research Needs

Although a number of limitations were cited in this section for the controlled use of immobilized enzymes in a variety of reactor configurations, a few additional considerations can be mentioned. The use of the simple one-substrate Briggs-Haldane or Michaelis-Menten equation as a kinetic model simply may be too inaccurate for a number of reactor design or analysis studies. Pore-diffusion rate-limiting conditions constitute an example. Another such case is the tubular flow-reactor in which the substrate undergoes very high conversion so that the ratio of $[E]_0$ to $[S]$ gets too large; processes in which enzyme reactors are used to remove trace amounts of impurities from waste streams may be especially subject to this modeling limitation. Since many enzyme reactions require two substrates, it may not be feasible to operate industrial enzyme-reactors with one substrate in large excess so that the more complex case of two-substrate kinetics will be needed. These are only a few but perhaps the more important of the reactor and model limitations that one can visualize. Although numerous others could be listed, there are other aspects of "Enzyme Engineering" which appear logically to have a higher research priority at present.

6. Applications and Future Directions

The past and present uses of enzymes have been discussed by Wingard (1972a) and DeBecze (1965); Guilbault (1966, 1968, 1970) has reviewed their uses in analytical chemistry. In the medical field several books are available on clinical enzymology.

Applications for immobilized enzymes are extremly easy to visualize but much more difficult to put into practical, economical practice. Melrose (1971) presented a list of applications of 27 different immobilized enzymes, including over 40 specific uses. However, in nearly every case these "applications" were reports in which the immobilized enzymes had been shown to function satisfactorily or even better than their soluble counterparts; in very few cases had any of these uses been put into practice. Goldman, Goldstein, and Katchalski (1971) also described numerous reports where immobilized enzymes had been tried in a number of research and development studies, but mentioned that most of these potential applications were only in their early stages of development. On the industrial scene the technology assessment report on enzymes by Rubin (1971) described a large number of potential applications of enzymes; but because of the short time available for preparing the report he went into the practicality of the matter very little. Commercially, a process using immobilized aminoacylase for the preparation of optically active aminoacids has been in operation in Japan for several years. In the United States, one or two commercial processes have been reported to utilize immobilized enzymes and three or four companies have processes under serious consideration for large-scale utilization. Mosbach (1971) has given a general discussion of potential applications of immobilized enzymes.

The application of immobilized enzymes in high-specificity electrodes for analytical chemistry likely will find practical and economical uses in the near future. Guilbault (1972) has described work in preparing a urease electrode for urea detection, an L-aminoacid-oxidase electrode and a similar one for D-aminoacids, an asparaginase electrode for asparagine, a rhodanase electrode to remove cyanide, and fluorescing electrodes for the detection of the enzyme cholinesterase. Clark (1972) has reported developing a polarographic electrode containing an alcohol oxidoreductase, capable of measuring methanol concentrations in the very low range of 1—25 nanomolar; it can be used also with ethanol and n-propanol but at reduced sensitivity. Hornby, Filippusson, and McDonald (1970) described the use of immobilized glucose oxidase for glucose analysis. Rechnitz and Llenado (1971) and Llenado and Rechnitz (1971) have developed a β-glucosidase electrode for amygdalin; and Baum and Ward (1971) have described a disc electrode for measuring cholinesterase activity.

A few recent medical studies and potential applications for immobilized enzymes include the use of matrix-supported (Hasselberger, 1970) and microencapsulated (Chang, 1971 b) asparaginase to reduce the asparagine level associated with certain cancers. Blatt and Kim (1971) used insulin bound to Sepharose to activate glycogen synthetase *in vitro* for studying the mechanism of action of insulin. Howell and Dupe (1971) observed that immobilized trypsin served as a useful tool for studying the activation of blood coagulation Factor X, while Falb and Grode (1971) have discussed use of immobilized enzymes for the preparation of artificial surfaces to be in contact with blood.

Several recent symposia on applications of enzymes included one at the Birmingham University (1971), another in France sponsored by the Commission Internationale des Industries Agricoles et Alimentaires (1972), one in late 1971 at the Battelle Institute in Columbus, Ohio, short ones in 1971 at Columbia University and the University of Pennsylvania, and undoubtedly others. The highly successful Engineering Foundation conference of Enzyme Engineering held in August 1971 in the United States is to have a sequel conference during the summer of 1973.

The future directions for enzyme engineering have been ably discussed by Edwards (1972). In addition to listing existing and potential uses of enzymes, Edwards has cited a number of problems and limitations that need to be resolved to bring about the practical utilization of these potential uses for enzymes.

In conclusion the potential for controlled use of enzymes both *in vitro* and *in vivo* and with enzymes immobilized for easy recovery and re-use looks very promising. However, the major restriction on the further development of "Enzyme Engineering" is the need to reduce more of these *potential* applications to actual practice and show that the enzyme route has a practical or economic advantage over other routes. Thus it seems appropriate to encourage the development of end-uses and specific, practical processes involving enzyme catalysts at this time with less (but certainly some) emphasis on sophisticated modeling and mathematical evaluations of reactor studies. The development of methods for intrinsic stabilization of intracellular enzymes in non-cellular environments, and the study and application of multi-enzyme reactor systems are two longer-range areas of research that would appear very fruitful.

Symbols

A_g	surface area per weight of catalyst,
B	void fraction,
C_s	salt concentration,
D	diffusion coefficient,
D_e	effective diffusion coefficient,
E	enzyme,
$[E]_0$	total enzyme concentration,
ES	enzyme-substrate complex,
e	membrane thickness,
$\mathfrak{E}$	effectiveness factor,
$\mathfrak{E}_1$	effectiveness factor for $K_m \gg [S]$,
F	Faraday constant,
f	fraction of substrate converted,
h_1	Thiele modulus for $K_m \gg [S]$,
h_0	Thiele modulus for $K_m \ll [S]$,
J	flux of substrate,
K	constant in Eq. (2),
K_m	true Michaelis constant $= \dfrac{k_2 + k_3}{k_1}$,
K'_m	apparent Michaelis constant,
k_n	rate constant for n-th reaction,
N	protein solubility,
P	product,
Q	flow rate,
R	gas constant,
R_0	particle radius,
r	radial distance,
$\bar{r}$	dimensionless radial distance $= r/R_0$,
S	substrate,
$[S]^*$	dimensionless substrate concentration $= [S]/[S]_0$,
$[\bar{S}]$	an average substrate concentration $= [S]_0 - [X]/2$,
t	time,
T	absolute temperature,
v	reaction velocity,
v_0	initial reaction velocity,
V_m	maximum reaction velocity $= k_3[E]_0$,
V'_m	apparent maximum velocity,
$\bar{v}$	an average velocity $= [X]/t$,

$[X]$	concentration of substrate converted to product,
y	thickness of stagnant diffusion layer,
W	axial distance,
$\mathfrak{z}$	electrical charge,
α	a dimensionless parameter $= [S]/K_m$,
β	a constant,
ψ	electrical potential,
ϱ_p	density of catalyst particle (solid and voids),
σ	a dimensionless parameter $= \dfrac{k_3[E]_0 e^2}{K_m D_e}$,
$[\ ]$	concentration,
$[\ ]_0$	initial concentration.

References

Axen, R., Ernback, S.: Europ. J. Biochem. **18**, 351 (1971).

Barman, T. E.: Enzyme Handbook, Vol. I] and II. Berlin-Heidelberg-New York: Springer 1969.

Baum, G., Ward, F. B.: Anal. Biochem. **42**, 487 (1971).

Belin, L., Hoborn, J., Falsen, E., Andre, J.: Lancet **1970 II,** 1153.

Bell, R. M., Koshland, D. E., Jr.: Science **172**, 1253 (1971).

Bernfield, P., Bieber, R. E., Watson, D. M.: Biochim. Biophys. Acta **191**, 570 (1969).

Birmingham University: Proc. One-Day Symp. Enzymes in Industry, Chemical Engineer., Vol. **22**; No. 2, 25—83 (1971).

Blatt, L. M., Kim, K.-H.: J. Biol. Chem. **246**, 4895 (1971).

Boguslaski, R. C., Janik, A. M.: Biochim. Biophys. Acta **250**, 266 (1971).

Bombaugh, K. J.: J. Chromatog. **53**, 27 (1970).

Bowski, L., Saini, R., Ryu, D. Y., Vieth, W. R.: Biotech. Bioeng. **13**, 641 (1971).

Bowski, L., Shah, P. M., Ryu, D. Y., Vieth, W. R.: In: Wingard, L. B., Jr. (Ed.): Enzyme Engineering, pp. 229—239, New York: John Wiley 1972.

Briggs, G. E., Haldane, J. B. S.: Biochem. J. **19**, 383 (1925).

Broun, G., Selegny, E., Minh, C. T., Thomas, D.: FEBS Letters **7**, 223 (1970).

Butterworth, T. A., Wang, D. I. C., Sinskey, A. J.: Biotech. Bioeng. **12**, 615 (1970).

Carbonell, R. G., Kostin, M. D.: Am. Inst. Chem. Eng. J. **18**, 1 (1972).

Ceska, M.: Europ. J. Biochem. **22**, 186 (1971).

Cha, S.: J. Biol. Chem. **245**, 4814 (1970).

Chang, T. M. S.: Science **146**, 524 (1964).

Chang, T. M. S.: Biochem. Biophys. Res. Commun. **44**, 1531 (1971 a).

Chang, T. M. S.: Nature **229**, 117 (1971 b).

Chang, T. M. S.: "Artificial Cells", monograph, Charles C. Thomas Publisher. Springfield, Ill., 1972.

Charm, S. E., Lai, C. J.: Biotech. Bioeng. **13**, 185 (1971).

Charm, S. E., Matteo, C. C., Carlson, R.: Anal. Biochem. **30**, 1 (1969).

Chock, P. B.: Biochemie **53**, 161 (1971).

Chrambach, A., Rodbard, D.: Science **173**, 440 (1971).

Clark, L. C., Jr.: In: Wingard, L. B., Jr. (Ed.): Enzyme Engineering, pp. 377—394, New York: John Wiley 1972.

Cleland, W. W.: Ann. Rev. Biochem. **36**, 77 (1967).

Cohen, R., Mire, M.: Eur. J. Biochem. **23**, 267 (1971).
Collier, R., Kohlhaw, G.: Anal. Biochem. **42**, 48 (1971).
Comm. Intern. Indust. Agric. Aliment.: Use of Enzymes in Agriculture and Food Industry, 24 rue de Teheran, Paris 75, France 1972.
Cuatrecasas, P.: In: Stark, G. R. (Ed.): Biochemical Aspects of Reactions on Solid Supports, pp. 79—109, New York: Academic Press 1971.
Cuatrecasas, P., Anfinsen, C. B.: Ann. Rev. Biochem. **40**, 259 (1971).
Darvey, I. G.: J. Theoret. Biol. **19**, 215 (1968).
Darvey, I. G.: J. Theoret. Biol. **25**, 109 (1969).
DeBecze, G. I.: In: Kirk, Othmer (Eds.): Encyclopedia of Chemical Technology, 2nd Ed., Vol. 8, pp. 173—230, New York: Interscience 1965.
Demain, A. L.: In: Wingart, L. B., Jr. (Ed.): Enzyme Engineering, pp. 21—32, New York: John Wiley & Sons 1972.
Denkewalter, R. G., Veber, D. F., Holly, F. W., Hirschmann, R.: J. Am. Chem. Soc. **91**, 502 (1969).
Dubos, R.: Science **173**, 259 (1971).
Dunnill, P., Dunnill, P. M., Boddy, A., Houldsworth, M., Lilly, M. D.: Biotech. Bioeng. **9**, 343 (1967 a).
Dunnill, P., Lilly, M. D.: Process Biochem. July (1967 b).
Dunnill, P., Lilly, M. D.: In: Wingard, L. B., Jr. (Ed.): Enzyme Engineering, pp. 97—113. New York: John Wiley 1972.
Edwards, V. H.: Advanc. Appl. Microbiol. **11**, 159 (1969).
Edwards, V. H.: In: Wingard, L. B., Jr. (Ed.): Enzyme Engineering, pp. 343—353. New York: John Wiley 1972.
Eigen, M., DeMaeyer, L.: In: Friess, S. L., Lewis, E. S., Weissberger, A. (Eds.): Investigations of Rates and Mechanisms of Reactions, Vol. 8, pp. 895—1054. New York: John Wiley 1963.
Elander, R. P.: In: Perlman, D. (Ed.): Fermentation Advances, pp. 89—114, New York: Academic Press 1969.
Epstein, W., Beckwith, J. R.: Ann. Rev. Biochem. **37**, 411 (1968).
Epton, R., McLaren, J. V., Thomas, T. H.: Biochem. J. **123**, 21P (1971).
Falb, R. A., Grode, G. A.: Federation Proc. **30**, 1688 (1971).
Filner, P., Wray, J. L., Varner, J. E.: Science **165**, 358 (1969).
Ford, J. R., Lambert, A. H., Cohen, W., Chambers, R. P.: In: Wingard, L. B., Jr. (Ed.): Enzyme Engineering, pp. 267—284. New York: John Wiley 1972.
Foster, P. R., Dunnill, P., Lilly, M. D.: Biotech. Bioeng. **13**, 713 (1971).
Franks, N. E.: In: Wingart, L. B., Jr. (Ed.): Enzyme Engineering, pp. 327—339. New York: John Wiley 1972.
Gabel, D., Vretblad, P., Axen, R., Porath, J.: Biochim. Biophys. Acta. **214**, 561 (1970).
Ghose, T. K., Kostick, J. A.: Biotech. Bioeng. **12**, 921 (1970).
Gibson, Q. H.: Ann. Rev. Biochem. **35**, 435 (1966).
Ginsburg, A., Stadtman, E. R.: Ann. Rev. Biochem. **39**, 429 (1970).
Glassmeyer, C. K., Ogle, J. D.: Biochemistry **10**, 786 (1971).
Goldman, R., Goldstein, L., Katchalski, E.: In: Stark, G. R., (Ed.): Biochemical Aspects of Reactions on Solid Supports, pp. 1—78. New York: Academic Press 1971.
Goldman, R., Katchalski, E.: J. Theoret. Biol. **32**, 243 (1971).
Goldman, R., Kedem, O., Katchalski, E.: Biochemistry **7**, 4518 (1968).
Goldman, R., Kedem, O., Katchalski, E.: Biochemistry **10**, 165 (1971).
Goldsmith, R. L.: Ind. Eng. Chem. Fundamentals **10**, 113 (1971).
Goldstein, L.: In: Perlman, D. (Ed.): Fermentation Advances, pp. 391—424. New York: Academic Press 1969.

Gregoriadis, G., Leathwood, P. D., Ryman, B. E.: FEBS Letters **14**, 95 (1971).

Guilbault, G. G.: Fund. Rev. Anal. Chem., Anal. Chem. **38**, 527R (1966).

Guilbault, G. G.: Ann. Rev. Anal. Chem., Anal. Chem. **40**, 459R (1968).

Guilbault, G. G.: Fund Rev. Anal. Chem., Anal. Chem. **42**, 334R (1970).

Guilbault, G. G.: In: Wingart, L. B., Jr. (Ed.): Enzyme Engineering, pp. 361—376. New York: John Wiley 1972.

Guilbault, G. G., Das, J.: Anal. Biochem. **33**, 341 (1970).

Gutte, B., Merrifield, R. B.: J. Am. Chem. Soc. **91**, 501 (1969).

Hasselberger, F. X., Brown, H. D., Chattopadhyay, S. K., Mather, A. N., Stasiw, R. O., Patel, A. B., Pennington, S. N.: Cancer Res. **30**, 2736 (1970).

Heineken, F. G., Tsuchiya, H. M., Aris, R.: Math. Biosciences **1**, 95 (1967).

Henri, V.: Lois generales de l'action des diastases. Paris: Hermann Publishers 1903.

Hess, B., Boiteux, A.: Ann. Rev. Biochem. **40**, 237 (1971).

Hicks, G. P., Updike, S. J.: Anal. Chem. **38**, 726 (1966).

Hornby, W. E., Filippusson: Biochim. Biophys. Acta **220**, 343 (1970).

Hornby, W. E., Filippusson, H., McDonald, A.: FEBS Letters **9**, 8 (1970).

Hornby, W. E., Lilly, M. D., Crook, E. M.: Biochem. J. **107**, 669 (1968).

Howell, R. M., Dupe, R. J.: Biochem. J. **123**, 11P (1971).

Ito, Y., Bowman, R. L.: Science **173**, 420 (1971).

IUB: Enzyme Nomenclature. New York: Elsevier Publ. Comp. (1964 Recommendations International Union of Biochemistry) 1965.

Jakoby, W. B.: In: Colowick, S. P., Kaplan, N. O. (Eds.): Methods in Enzymology, Vol. 22. New York: Academic Press 1971.

Johnston, M. M., Diven, W. F.: J. Theoret. Biol. **25**, 331 (1969).

Julliard, J. H., Godinot, C., Gautheron, D. C.: FEBS Letters **14**, 185 (1971).

Katchalski, E., Sela, M., Silman, H. I., Berger, A.: In: Neurath, H. (Ed.): The Proteins, Vol. 2, pp. 405—540, 2nd ed., 1964.

Katchalski, E., Silman, I., Goldman, R.: In: Nord, F. F. (Ed.): Advances in Enzymology, Vol. 34, pp. 445—536. New York: Interscience (Wiley) 1971.

Kaufman, B. T., Pierce, J. V.: Biochem. Biophys. Res. Commun. **44**, 608 (1971).

Kay, G., Lilly, M. D., Sharp, A. K., Wilson, R. J. H.: Nature **217**, 641 (1968).

Khidekel, M. L.: In: Balandin, A. A. (Ed.): Scientific Selection of Catalysts, pp. 231—238. Hartford, Conn.: Daniel Davey & Co. 1968.

Klotz, I. M., Royer, G. P., Searpa, I. S.: Proc. Nat. Acad. Sci. **68**, 263 (1971).

Kobayaski, T., Moo-Young, M.: Biotech. Bioeng. **13**, 893 (1971).

Koch, A. I., Coffman, R.: Biotech. Bioeng. **13**, 651 (1970).

Kozinski, A. A., Lightfoot, E. N.: Am. Inst. Chem. Engrs. J. **17**, 81 (1971).

Kunitake, T., Shinkai, S.: J. Am. Chem. Soc. **93**, 4256 (1971).

Lampen, J. O.: In: Wingard, L. B., Jr. (Ed.): Enzyme Engineering, pp. 37—41. New York: John Wiley 1972.

Larsson, P. O., Mosbach, K.: Biotech. Bioeng. **13**, 393 (1971).

Laurent, T. C.: European. J. Biochem. **21**, 498 (1971).

Lee, H. J., Wilson, I. B.: Biochim. Biophys. Acta **242**, 519 (1971).

Lee, J. C.: Biochim. Biophys. Acta **235**, 435 (1971).

Li, N. N.: Am. Inst. Chem. Engrs. J. **17**, 459 (1971).

Lichtenstein, L. M., et al.: J. Allergy **47**, 53 (1971).

Lilly, M. D.: Biotech. Bioeng. **13**, 589 (1971).

Lilly, M. D., Dunnill, P.: In: Wingard, L. B., Jr. (Ed.): Enzyme Engineering, pp. 221—227. New York: John Wiley 1972.

Lilly, M. D., Hornby, W. E., Crook, E. M.: Biochem. J. **100**, 718 (1966).

Lilly, M. D., Sharp, A. K.: The Chemical Engineer, No. 215, January-February, CE12—CE18. (publ. by Institution of Chemical Engrs., England) 1968.

Lindsey, A. S.: J. Macromol. Sci.-Rev. Macromol. Chem. C3, 1 (1969).
Line, W. F., Kwong, A., Weetall, H. H.: Biochim. Biophys. Acta 242, 194 (1971).
Llenado, R. A., Rechnitz, G. A.: Anal. Chem. 43, 1457 (1971).
Lowe, C. R., Dean, P. D. G.: FEBS Letters 14, 313 (1971).
Marcus, A.: Ann. Rev. Plant Physiol. 22, 313 (1971).
Marglin, A., Merrifield, R. B.: Ann. Rev. Biochem. 39, 841 (1970).
Marshall, D. L., Walter, J. L., Falb, R. D.: In: Wingard, L. B., Jr. (Ed.): Enzyme Engi-
    neering, pp. 195—209. New York: John Wiley 1972.
Mattiasson, B., Mosbach, K.: Biochim. Biophys. Acta 235, 253 (1971).
McKenzie, H. A., Ralston, G. B.: Experientia 27, 617 (1971).
McLaren, A. D., Packer, L.: Adv. Enzymol. 33, 245 (1970).
Meighen, E. A., Pigiet, V., Schachman, H. K.: Proc. Nat. Acad. Sci. 65, 234 (1970).
Melrose, G. J. H.: Rev. Pure Appl. Chem. 21, 83 (1971).
Merrifield, R. B.: J. Am. Chem. Soc. 85, 2149 (1963).
Merrifield, R. B., Stewart, J. M., Jernberg, N.: Anal. Chem. 38, 1905 (1966).
Messing, R. A.: Enzymologia 39, 12 (1970).
Michaels, A. S.: In: Perry, E. S. (Ed.): Progress in Separation and Purification, Vol. 1,
    pp. 297—334. New York: John Wiley 1968.
Michaelis, L., Menten, M. L.: Biochem. Z. 49, 333 (1913).
Monsan, P., Durand, G.: FEBS Letters 16, 39 (1971).
Mosbach, K.: Sci. Am. 224, 26 March (1971).
Mosbach, K., Larsson, P.-O.: Biotech. Bioeng. 12, 19 (1970).
Mosbach, K., Mattiasson, B.: Acta Chem. Scand. 24, 2093 (1970).
Neurath, A. R., Weetall, H. H.: FEBS Letters 8, 253 (1970).
Newbold, P. C. H., Harding, N. G. L.: Biochem. J. 124, 1 (1971).
Nikolaev, L. A.: Russ. Chem. Rev. 33, 275 (1964).
Ollis, D. F.: Diffusion Influences in Denaturable Insolubilized Enzyme Catalysts,
    64th Ann. Mtg. Am. Inst. Chem. Engrs., December, San Francisco, Calif. 1971.
O'Neill, S. P.: Biotech. Bioeng. 13, 493 (1971).
O'Neill, S. P., Dunnill, P., Lilly, M. D.: Biotech. Bioeng. 13, 337 (1971).
O'Neill, S. P., Lilly, M. D., Rowe, P. N.: Chem. Eng. Sci. 26, 173 (1971).
O'Neill, S. P., Wykes, J. R., Dunnill, P., Lilly, M. D.: Biotech. Bioeng. 13, 319 (1971).
Otto, N. C., Quinn, J. A.: Chem. Eng. Sci. 26, 949 (1971).
Otto, N. C., Quinn, J. A.: In: Wingard, L. B., Jr. (Ed.): Enzyme Engineering,
    pp. 311—321. New York: 1972.
Overberger, C. G., Salamone, J. C., Yaroslavsky, S.: Pure Appl. Chem. 15, 453 (1967).
Palit, S. R.: Sci. Cult. 34, 37 (1968).
Pardee, A. B.: In: Perlman, D. (Ed.): Fermentation Advances, pp. 3—14. New York:
    Academic Press. 1969.
Pocklington, T., Jeffery, J.: Biochem. J. 112, 331 (1969).
Porath, J.: In: Wingard, L. B., Jr. (Ed.): Enzyme Engineering, pp. 145—166. New
    York: John Wiley 1972.
Porter, M. C.: In: Wingart, L. B., Jr. (Ed.): Enzyme Engineering, pp. 115—144. New
    York: John Wiley 1972.
Rechnitz, G. A., Llenado, R.: Anal. Chem. 43, 283 (1971).
Reed, L. J., Cox, D. J.: In: Boyer, P. D. (Ed.): The Enzymes, Vol. 1, 3rd ed., pp. 213—
    240. New York: Academic Press 1970.
Rhoads, D. G., Acks, M. J., Peterson, L., Garfinkel, D.: Computers Biomed. Res. 2,
    45 (1968a).
Rhoads, D. G., Pring, M.: J. Theoret. Biol. 20, 297 (1968).
Robinson, P. J., Dunnill, P., Lilly, M. D.: Biochim. Biophys. Acta 242, 659 (1971).
Robinson, N. C., Tye, R. W., Neurath, H., Walsh, K. A.: Biochem. 10, 2743 (1971a).

Rony, P. R.: Biotech. Bioeng. **13**, 431 (1971).

Rubin, D. H.: "Technology Assessment Series: Enzymes (Industrial)", National Technical Information Service, PB 202778—04, Springfield, Va. 1971.

Sano, S., Tokunaga, R., Kun, K. A.: Biochim. Biophys. Acta **244**, 201 (1971).

Savageau, M. A.: J. Theoret. Biol. **25**, 365 (1969a).

Savageau, M. A.: J. Theoret. Biol. **25**, 370 (1969b).

Schimke, R. T., Doyle, D.: Ann. Rev. Biochem. **39**, 929 (1970).

Schramm, M.: Ann. Rev. Biochem. **36**, 307 (1967).

Schurr, J. M.: Biophys. J. **10**, 717 (1970).

Schwencke, J., Farris, G., Rojas, M.: European. J. Biochem. **21**, 137 (1971).

Schwert, G. W.: J. Biol. Chem. **244**, 1278 (1969).

Selegny, E., Broun, G., Thomas, D.: Physiol. Veg. **9**, 25 (1971).

Selegny, E., Kernevez, J.-P., Broun, G., Thomas, D.: Physiol. Veg. **9**, 51 (1971).

Sharp, A. K., Kay, G., Lilly, M. D.: Biotech. Bioeng. **11**, 363 (1969).

Sheehan, J. C.: Proc. Conf. Struct. React. DFP Sensitive Enzymes, Stockholm 1966; Heilbron, E. (Ed.): pp. 59—68; Forskningsanstalt, Stockholm, Sweden 1967.

Shrier, A. L.: In: Wingard, L. B., Jr., Enzyme Engineering, pp. 323—326. New York: John Wiley 1972.

Smiley, K.: Biotech. Bioeng. **13**, 309 (1971).

Smith, W.: Bull. Math. Biophys. **33**, 97 (1971).

Sober, H. A., Hartley, R. W., Jr., Caroll, W. R., Peterson, E. A.: In: Neurath, H. (Ed.): The Proteins, Vol. 3, 2nd ed., pp. 2—98. New York: Academic Press 1964.

Staff, P. J.: J. Theoret. Biol. **27**, 221 (1970).

Stesina, L. N., Akopyan, Zh. I., Gorkin, V. Z.: FEBS Letters **16**, 349 (1971).

Stewart, J. M., Young, J. D.: Solid Phase Peptide Synthesis. San Francisco, Calif.: W. J. Freeman & Co. 1969.

Sundaram, P. V., Pye, E. K.: In: Wingard, L. B., Jr. (Ed.): Enzyme Engineering, pp. 15—18. New York: John Wiley (Report of ad-Hoc Committee) 1972.

Swanljung, P.: Anal. Biochem. **43**, 382 (1971).

Tanner, R. D.: Identification, Hysteresis, and Discrimination in Enzyme Kinetic Models, 69th Nat'l. Mtg. Am. Inst. Chem. Engrs., Cincinnati, Ohio 1971.

Tanner, R.: Ind. Eng. Chem. Fundamentals **11**, 1 (1972).

Teipel, J. W., Koshland, D. E., Jr.: Biochem. **10**, 792 (1971), **10**, 798 (1971).

Telling, R. C., Radlett, P. J.: In: Perlman, D. (Ed.): Advances in Applied Microbiology, Vol. 13, pp. 91—119. New York: Academic Press 1970.

Uren, J. R.: Biochim. Biophys. Acta **236**, 67 (1971).

Vergonet, G., Berendsen, H. J. C.: J. Theoret. Biol. **28**, 155 (1970).

Vermeulen, T., Nady, L., Krochta, J. M., Ravoo, E., Howery, D.: Ind. Eng. Chem. Process. Desig. Develop. **10**, 91 (1971).

Vieth, W. R., Gilbert, S. G., Wang, S. S.: In: Wingard, L. B., Jr. (Ed.): Enzyme Engineering, pp. 285—297. New York: John Wiley 1972.

Vretblad, P., Axen, R.: FEBS Letters 18 (1971).

Wang, J. H.: J. Am. Chem. Soc. **77**, 4715 (1955).

Weber, K., Kuter, D. J.: J. Biol. Chem. **246**, 4504 (1971).

Weetall, H. H.: Biochim. Biophys. Acta **212**, 1 (1970).

Weetall, H. H., Baum, G.: Biotech. Bioeng. **12**, 399 (1970).

Weetall, H. H., Havewala, N. B.: In: Wingard, L. B., Jr. (Ed.): Enzyme Engineering, pp. 241—266. New York: John Wiley 1972.

Weetall, H. H., Hersh, L. S.: Biochim. Biophys. Acta **206**, 54 (1970).

Weibel, M. K., Bright, H. J.: Biochem. J. **124**, 801 (1971).

Weibel, M. K., Doyle, E. R., Humphrey, A. E., Bright, H. H.: In: Wingard, L. B., Jr. (Ed.): Enzyme Engineering, pp. 167—171. New York: John Wiley 1972.

Weibel, M. K., Humphrey, A. E.: 64th Annual Meeting Am. Inst. Chem. Engrs., December 1971, San Francisco, Calif 1971.

Weibel, M. K., Weetall, H. H., Bright, H. J.: Biochem. Biophys. Res. Commun. **44**, 347 (1971).

Weiker, H. J., Johannes, K. J., Hess. B.: FEBS Letters **8**, 178 (1970).

Wheeler, A.: In: Advances in Catalysis, Vol. 3, pp. 249—327. New York: Academic Press 1951.

Wilchek, M., Bocchini, V., Becker, M., Givol, D.: Biochem. **10**, 2828 (1971).

Wilson, R. J. H., Kay, G., Lilly, M. D.: Biochem. J. **109**, 137 (1968).

Winchester, B. G., Caffrey, M., Robinson, D.: Biochem. J. **121**, 161 (1971).

Wingard, L. B., Jr.: In: Wingard, L. B., Jr. (Ed.): Enzyme Engineering, pp. 3—13. New York: John Wiley (Spec. Suppl. Biotech. Bioeng.) 1972a.

Wingard, L. B., Jr.: (Ed.): Enzyme Engineering. New York: John Wiley (special supplement of Biotech. Bioeng.) 1972b.

Wingard, L. B., Jr., Finn, R. K.: Chem. Eng. Progr. Symp. Ser. **62**, (No. 69), 30 (1966).

Wingard, L. B., Jr., Lui, C. C.: Proc. 8th Intl. Conf. Med. Biol. Eng., July, 1969, Chicago, p. 26 (1969).

Wingard, L. B., Jr., Liu, C. C., Nagda, N. L.: Biotech. Bioeng. **13**, 629 (1971).

Yamamoto, H., Noguchi, J.: J. Biochem. **67**, 103 (1970).

Zaborsky, O.: In: Wingard, L. B., Jr. (Ed.): Enzyme Engineering, pp. 211—217. New York: John Wiley 1972.

L. B. Wingart, Jr., Ph. D.,
Research Associate Professor in Pharmacology
and Chemical Engineering
Department of Pharmacology
University of Pittsburgh
Scaife Hall, School of Medicine
Pittsburgh, PA 15213, USA.

CHAPTER 2

Application of Computers in Biochemical Engineering

L. K. Nyiri

With 18 Figures

Contents

1. Introduction

Developments in molecular biology have resulted in excellent genetic techniques for selection of strains for fermentation processes. However, the responses of these strains to environmental factors are still to be fully understood. Recently there has been tremendous development of instrumentation to detect and control the environmental factors and this advance is reducing the gap between the knowledge about the genetics and the metabolic behavior of the cells under various environmental conditions.

As the number of environmental sensors has increased, the need for better and more efficient methods of data collection and interpretation has become a pressing problem. In addition it has opened the door to the possibility of using the recently developed control techniques for the control of environment for fermentation processes. These developments are making the present fermentation industry a fruitful field for computer application.

In the past years there has been an amazing proliferation of computer applications in the petroleum and chemical industries. This has occurred because computers have proven useful in handling large masses of data with high operational speed. This successful operation of computers both in scientific and industrial fields predicted their applicability for biochemical engineering purposes too. As usual, here computer utilisation is also based on its high data-processing capability. Hence, ultimately the computer will find use to define and create the optimum environment for a desired fermentation performance.

The recent dramatic reduction in computer installation costs also stimulates the increasing utilisation of computers in many places related to biochemical engineering. Experience of previous computer users indicates that, as an advantage of computer utilisation, the time lag between the phenomenon and its observation will be reduced and hence the scope of observation will widen. This will transform significantly the methods of planning experiments and processing data.

With this background in mind this paper was written to analyse the status, future trends and limitations of computer utilisation in biochemical engineering.

2. Literature Survey

Since this paper is among the first to appear on the subject it includes a brief literature survey on the related works. Here the following subjects will be covered: 1. model construction and process analysis with computers in biochemical engineering, 2. direct application of computers in fermentation processes and 3. miscellaneous application of computers[1].

---

[1] The literature survey ended by May 31, 1971.

a) Modeling and Process Analysis with Computers

Analysis of biological processes by means of model-systems resulted in increased insight into the functional characteristics and basic mechanism of living systems (Heinmets, 1969, Garfinkel *et al.*, 1970).

A mathematical model of microbial cell-growth was formulated by Monod (1949). Kinetic patterns of various fermentation systems were defined by Gaden (1955) and Deindoerfer (1960).

A mathematical relationship between growth and product formation was formulated by Luedeking and Piret (1959). All of these works have already been reviewed in the book written by Aiba and his coworkers (1965).

The need for more accurate representation of microbial cell growth produced more detailed mathematical models for microbial kinetics. According to the literature there are two types of models related to microbial life cycles and biochemistry, namely:

1. Unstructured models in which a uniformly distributed cell mass is considered along with the relationship between the biomass and product formation. Unstructured models are compiled in Table 4.

2. Structured models where the cell structure is also considered. In this model the total nitrogen content is differentiated (DNA, RNA, protein); furthermore there is a possibility of considering enzyme activities and metabolic pathways.

From a methodological point of view models have been analysed in two general ways:

A. Simulation of dynamic behavior of the model when the environmental conditions are theoretically optimal. This type of analysis permits consideration of relationships between the consecutive steps of the process.

B. Analysis of the dynamic behavior of the model with specific respect to the effect of environment on the particular process (optimisation).

Many models having become more complex and less manageable by hand calculations have been studied with computers (Table 1). It is noteworthy, however, that every kind of model formulated in mathematical equations is the potential subject of computerised analysis.

Unstructured Models

Shu (1961) made an unstructured model for the product formation in microbiological processes. An interesting feature of this model suggests consideration of age of cells and distribution of cells of different ages in

Table 1. Application of computers in analysis of cell growth and metabolism

Subjects	Related to	Computer	Language	Ref.
Batch fermentation of gluconic acid	Simulation	IBM 7040/I	FORTRAN	Koga et al. (1967)
Growth dynamics in chemostat	Simulation	analog		Andrews (1968)
Mathematical representation of batch culture data	Data analysis		WATFOR	Edwards and Wilke (1968)
Calculate and correlate heat evolution of cells	Data analysis		Project MCA	Cooney et al. (1969)
Prediction of product formation	Optimisation	IBM 360	FORTRAN IV	Gyllenberg et al. (1969)
Oxygen transfer rate measured by enzymatic reaction	Simulation	IBM 7094 + Philbrick analog	FORTRAN	Hsieh et al. (1969)
Model of biological oxidation process	Data fitting to experimental results	IBM 360/50	FORTRAN	Naito et al. (1969)
Flow models in multistage fermenter	Simulation	IBM 360/65	FORTRAN IV	Prokop et al. (1969)
Effect of mixing on growth	Simulation	IBM 360	RKGS	Tsai et al. (1969)
Cell growth in chemostat in presence of growth inhibitors	Simulation	HITACHI analog		Yano and Koga (1969)
Effect of high substrate concentration on growth	Fitting functions to experimental data	IBM 360/65	WATFOR	Edwards (1970)
Temperature profiles for penicillin fermentation	Optimisation	IBM 360	FORTRAN IV	Constantinides et al. (1970a, 1970b)
Acetate concentration and growth of yeasts	Simulation. Data fitting to experimental results	IBM 360/65	FORTRAN	Edwards et al. (1970)
Cell growth on hydrocarbons	Simulation	IBM 360	CSMP	Erickson et al. (1970)
Analysis of reactor performance in multi-stage tower fermenter	Simulation	IBM 7094	FORTRAN (MIMIC)	Falch and Gaden (1970)
Enzymatic conversion of penicillin G into 6-APA	Optimisation	IBM 360/75	FORTRAN IV	Ho and Humphrey (1970)

Table 1 (continued)

Subjects	Related to	Computer	Language	Ref.
Cell differentiation and product formation	Fit experimental data to simulated values			Megee et al. (1970)
Effect of ethanol on growth in continuous culture	Simulation	analog		Zines and Rogers (1970)
Dynamic chemostat model	Simulation. Fit data to ex-perimental results	analog		Young (1970)
Dynamic chemostat model	Simulation	analog		Young et al. (1970)
Fermentation process game	Simulation	IBM 360	FORTRAN IV	Bungay (1971)
Reproduction of RNA phage	Simulation	IBM 360/65		Chen et al. (1971)
Steady state model for activated sludge	Calculation of constants	IBM 1620		Ramanathan and Gaudy (1971)

the growth stages. A generalised logistic equation was constructed by Edwards and Wilke (1968) for mathematical representation of batch culture data. A computer trial to fit growth-associated product formation to growth in the case of lactic acid formation was successful.

Kono and Asai (1969a, 1969b) formulated a general mathematical model useful for a variety of fermentation process. Considering the critical cell concentrations at the boundary points of the individual growth phases, induction, transient, exponential growth and declining growth and additional phases were distinguished. The value of a constant in the general model called the "consumption activity coefficient" (apparent coefficient of growth activity) was determined for each growth phase. The versatile equations obtained were applied to analyse the conditions in batch and continuous fermentations (Kono and Asai, 1969a, 1969b). An attempt was made to analyse the population characteristics of con-taminants and mutants during continuous growth. The original equa-tions were later extended to obtain general expressions for the analysis of relationships between cell growth and product formation.

Kinetic analyses of lactic acid, sorbose, citric acid, and novobiocin fer-mentations using these equations proved that the experimental results are in good agreement with the theoretical predictions. More recently the kinetics of diauxic growth and the effect of operating conditions on cell mass production were successfully investigated by the use of these new equations (Kono and Asai, 1971a, 1971b).

The assistance of the computer was requested for simulation purposes of dynamic behavior of cell growth in the chemostat (Andrews, 1968), in the presence of a controlled amount of substrate (Edwards et al., 1970) and in the case when the substrate concentration was at an inhibitory level (Edwards, 1970; Yano and Koga, 1969).

Ramkrishna et al. (1966) produced a model on the basic assumption that an inhibitor-type compound is formed during the growth which influences growth and produces no viable cell mass. Zines and Rogers (1970) stated that when the system contains process inhibitors the transient response of the growth rate can be best described by second order equations.

On the basis of experiments with L. delbruckii in dialysis culture Friedman and Gaden (1970) modified the Luedeking-Piret model interpreting the lactate inhibition effect on growth.

Models have been constructed considering a Michaelis-Menten relationship between substrate concentration and growth in order to simulate the growth conditions in culture broths with two liquid phases [Erickson and Humphrey, 1969a, 1969b; Erickson et al. (1970)]. These experiments proved to be useful for better understanding of microbial growth on hydrocarbons.

A system engineering (block diagram) approach has resulted in dynamic mathematical models of chemostats (Young, 1970; Young et al., 1970). Here, analog computer simulations revealed that the dynamic relationships between μ and the control variables during transient conditions of growth can be determined from experimental pulse testing to get frequency response data. Regarding its importance this analytical approach will be studied in more details in another part of this paper.

A mathematical model was constructed by Chen et al. (1971) to study the kinetics of RNA phage production in batch culture. The model considered the basic steps in the phage-bacterial cell interactions. A set of differential equations describing the rates of change of cell concentration and phages in various stages of phage infection was solved by computer simulation, using Hamming's modified predictor-corrector method. The results of the simulation were compared with experimental data obtained from the production of MS 2—RNA phage on male type E. coli C300 as host cell. Discrepancies were found between the experimental and simulation data. Since microbial cells show different physiological characteristics during their growth there is a need for further refinement of the proposed model taking the absorptivity of E. coli and the physiological status of the cells into consideration.

Structured Models

A structured model for analog computer simulation of cell growth was recently constructed by Heinmets (1969). The model contains 19 differential equations and 46 constants (i.e. 23 constants of reversible enzymatic reactions) and is a general description from the molecular biology point of view of cell growth and metabolic activity.

Koga and his coworkers (1969) described a molecular kinetic theory in the form of a simple model. Formation and availability of ATP in anaerobic and aerobic systems are considered with specific respect to the pacemaking step during metabolism. Domain maps constructed on the basis of equations can be helpful in identification of the actual conditions of metabolism.

An enzyme kinetic model with ability to incorporate control variables for biochemical engineering was constructed by Tanner (1970). The model considers the reproduction and participation of ribosomes in the synthesis of various proteins and the role of enzymes in product formation. Following the limiting-substrate-concentration scheme differential equations were created for a sequence of events in the biosynthetic machinery starting with the formation of ribosomal RNA on DNA templates and leading to product formation through an adequate enzyme system. The test of the derived model with data from a gluconic acid fermentation demonstrated that the model may fit the data. Some explanations were found for the unusual hysteresis-type behavior of the rate of gluconic acid formation as a function of gluconolactone concentration and time. Nevertheless, computer simulation of Henri's enzyme model revealed that the hysteresis phenomenon is due to the occurrence of true steady state during the enzymatic reaction (Nyiri and Tóth, 1971). A mathematical model for growth of *Aspergillus awamori* took explicit account of cell growth and product formation (Megee *et al.*, 1970). The model considers the cell structure too. Five states of cell differentiation were distinguished: 1. active growth of tip, 2. appearance of growth-associated product (P_1) and 3.—5. appearance of non-growth-associated products (P_2, P_3, P_4). Formation of RNA, DNA, cell-protein and five enzymes in each differentiation state were then examined during growth, differentiation, branching of hyphae, dormancy because of lack of substrate, assimilation and product formation.

Since at least two separate states of growth are present during non-growth-associated product formation, a suitable control of the environment in the first growth stage may result in the accumulation of biomass, then in the next stage (production) the change of environmental conditions can serve for increased product formation. Data obtained by model simulation were in accordance with those described by Pirt and Callow (1960) as well as Pirt and Righelato (1967).

The necessity is emphasised of defining explicitly the growth and product formation parameters. The environmental optima for each growth stage can be found with the knowledge of distribution of differentiation states. These conclusions *per se* argue for detailed pilot-plant scale operations using highly instrumented fermentors and computerised, on-line data analysis to define the environmental optima from stage to stage.

Optimisation

Trials to find optimal values of environmental variables for fermentation processes indicate that mathematical models of growth and product formation can be useful for optimisation purposes if 1. a relatively correct model for the particular process is available (Koga *et al.*, 1967) and/ or 2. there is a possibility to apply a general model for a specific process using a set of experimental data for parameter evaluations (Ho and Humphrey, 1970; Constantinides *et al.*, 1970a, 1970b)

All of these approaches, however, indicate the extreme complexity of fermentation models and their optimisation procedures. Three factors are involved here, namely 1. inadequate knowledge of the intracellular biochemical processes as well as of the effects of external environmental conditions on these internal processes, 2. the number of interrelating control and state variables, and 3. the complexity of the objective function. These factors indicate the necessity of stepwise construction and computer analysis of simple and realistic models with a number of validity tests (Yamashita *et al.*, 1969; Constantinides *et al.*, 1970a, 1970b). Several mathematical procedures have been tried to find optimal process variables and predict the fate of a fermentation. To find the solution for 6 APA yield in the enzymatic conversion of benzylpenicillin to 6-aminopenicillanic acid Ho and Humphrey (1970) successfully used the common optimisation technique known as the "gradient method functional space" (Storey and Rosenbrook, 1964). In order to define the optimum temperature profiles for penicillin fermentations Constantinides and his coworkers (1970a, 1970b) used a non-linear regression technique for parameter evaluation and Pontryagin's continuous maximum principle technique for optimisation.

Gyllenberg *et al.* (1969) employed a matrix method to analyze the prediction of yield during fermentation. Data collected from 11 *Streptomyces aureofaciens* pilot-plant fermentations yielding tetracycline indicate that the trend of product formation can be predicted on the basis of accumulated experience and knowing the prevailing conditions. It is interesting to note that in this approach the succession of classes of deviations from the trend fitted the Markov chain and the top product yield could be predicted with sufficient accuracy.

Naito *et al.* (1969) analyzed a general model of biochemical oxidation with specific regard to the enzyme activity of the microbial cells involved in the process. The mathematical technique for estimation of parameters included a simple and useful method to minimise the sum of squares of deviations between the experimental and calculated data.

Furthermore, computers were used during the design and analysis of reactor performance in the case of a multistage continuous tower fermentor (Prokop *et al.*, 1969; Falch and Gaden, 1970). Here search procedures were applied to test the fit between the measured response curve and different calculated values which were obtained, solving the differential equations by inverse transformation of the Laplace solution after numerical determination of the characteristic roots.

More recently, Bungay reported the construction of a fermentation game (1971). The game from a computational point of view is considered as a case simulation in which the model has a determined human component. This approach is considered to be of importance from the viewpoint of education in biochemical engineering.

b) Direct Application of Computers in Fermentation Processes

The idea of using computers to control fermentation was first proposed by Fuld (1960). Since then accumulation of knowledge of fermentation processes has resulted in the enumeration of a set of control and state variables on different environmental levels (Table 2). Manipula-

Table 2. Control and state variables of fermentation processes on different environmental levels[a]

Physical	Chemical	Molecular biological	Biological
		Environments	
Temperature	pH of the broth	DNA level	Contamination
Vessel pressure	Redox potential	RNA level	Mutation
Power input	Dissolved oxygen	Total protein	
Agitation speed	Dissolved carbon dioxide	Specific protein (enzyme activity)	
Gas flow rates	Carbohydrate level		
Foaming	Nitrogen source level		
Liquid feed rates	Precursor level		
Viscosity	Mineral ion level		
Liquid volume			

[a] After data of Aiba *et al.* (1965) and Humphrey (1971).

Table 3. Direct application of computers in fermentation processes

Purpose	Scale	Function	Computer	Language	Ref.
Penicillin batch fermentation	Industrial	DDC	ARCH-102	Machine	Grayson (1969)
Glutamic acid batch fermentation	Industrial	DDC	YODIC-500		Yamashita *et al.* (1969)
General	Pilot-plant	data logging data analysis process control	PDP-11/20	Assembly	Nyiri (1971a)

ting these variables should result in adequate conditions for desired microbial growth and metabolism.

Table 3 presents the available information on the application of computers in fermentation processes.

As was reported by Grayson (1969) 36 fermentation vessels for batch production of penicillin are under Direct Digital Control. The computer is put in control of temperature and pH of the culture broth as well as the airflow rate and the addition of antifoam. In addition to the teletype printout a four-digit electronic system displays the values associated with any of the 114 control loops.

Yamashita and his coworkers (1968) described a concept of application of computers for control of large-scale glutamic acid fermentations. The plant was first manipulated with a relay-type sequence control system (Mori and Yamashita, 1967). A digital computer was then developed to replace the logic circuit of the sequence control system. A further development resulted in the application of this digital computer for control of several control variables, first in pilot-plant [Hoshi *et al.* (1967) then in large-scale units (Yamashita, 1967)]. Direct Digital Control was used which, akin to cases in the chemical industries (Cotter, 1969; Henri, 1970), was found to be adequate in replacing the conventional analog controllers to accomplish complex control performance. The computer monitored the following control variables: temperature of culture broth, vessel pressure, pH of the broth, airflow rate, addition of antifoam agent. As a general feature, in both cases the application of the computer was process-control-oriented; however, a need for methods to perform self-optimisation of the process was emphasised.

A pilot-plant scale fermentor-instrument-computer-logic complex was described by Nyiri (1971a, 1971b). Although the computer can perform control on all variables which influence the physical and chemical environment (except dissolved CO_2 concentration) the function of the computer is oriented towards data acquisition and data analysis, having a flexible, extendable, open-ended program structure for these purposes. The main goal of this fermentor-computer system is to increase the number of on-line observations of fermentation processes in order to find suitable algorithms for coordinated process control.

c) Miscellaneous Applications of the Computer in Fermentation Industry

Computer application has many possibilities where rapidity of calculation and data storage capacity are primary requirements. This is the case when the potency of biologically active materials is measured for product quality control purposes. Burns (1971) reported a device for direct measurement of antibiotic inhibition zones on agar plates. An on-line operating IBM 1800 computer monitors the function of this device. The printout is the actual antibiotic concentration of the sample with an evaluation of mean values and deviations from the mean. The advantage of this device is its relatively high working speed (3 measurements and printout/second) and the ability to store data for any kind of data retrieval.

Tuffile and Pinho (1970) used an IBM 1130 computer and a suitable program to study the changes of viscosity values during streptomycin fermentation. Although the operation was off-line, i.e. the computer was not directly interfaced with the instruments on the fermentor, this report is among the firsts where computer utilisation is mentioned to keep track of changes of fermentation characteristics.

Commonly, computers are used in the fermentation industry to produce daily, weekly and monthly reports on the production (Yamashita *et al.*, 1969). This application undoubtedly improves cooperation between departments during production and helps management in decision.

3. Methodological Considerations of Computer Utilisation

Problems related to biochemical engineering can be managed by computers essentially in two ways. Studies on theories related to the dynamic behavior of cell growth, product formation, computation for process optimisation and for reactor performance are executed by *off-line* computer operation (Fig. 1). In this case the computer has no actual contact with the process. On the other hand, the computer can perform *on-line* operation (Fig. 2). Here the sensors are connected with the computer through a suitable interface; sensor signals are evaluated on a real-time basis using a process-oriented computer program. The obtained data can be stored or printed out as well as used for process evaluation and/or

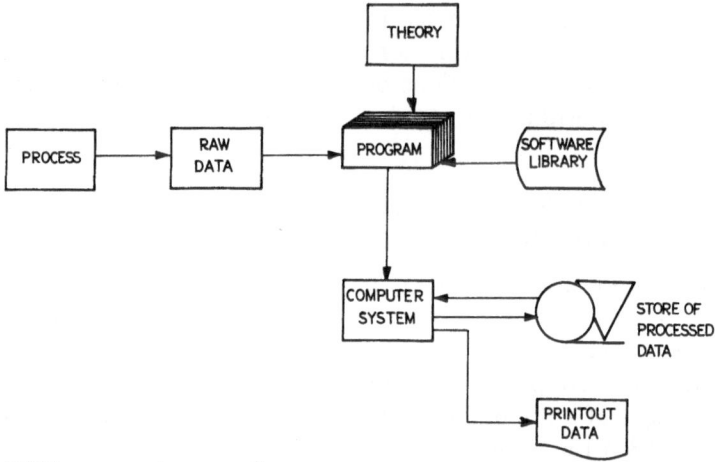

Fig. 1. Off-line computer operation

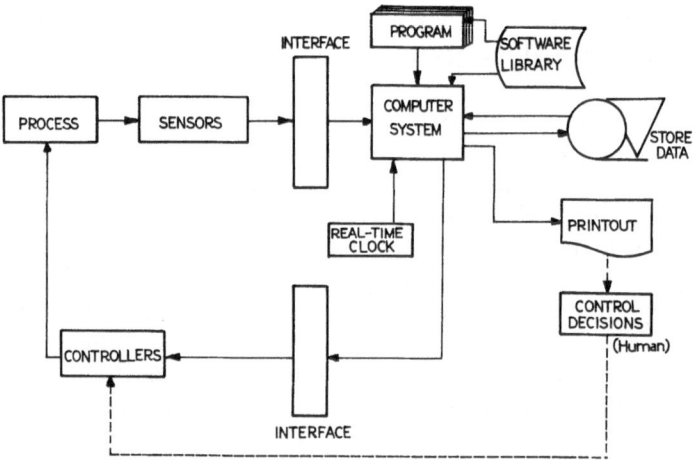

Fig. 2. On-line computer operation

control. The importance of this on-line, real-time "cooperation" between the fermentor and computer is emphasized by the increased need for exact, up-to-the-minute knowledge on the status of the fermentation which is the condition of flexible process control.

Both methods of data processing need specific strategy to assure the best possible utilisation of the computer. Nevertheless, the strategies related to both off-line and on-line operations can be reduced to the basic task of organisation and collection of data for computer use as well as to the construction of suitable programs to operate the computers.

a) Off-line Operation

Off-line utilisation of the computer is and presumably will be the check-point of dynamic behavior of more and more complex models of cell-growth and metabolic activity, as well as studies on the effects of environment on growth and cell metabolism, reactor performance and scale-up procedures.

Physical and Organisation Requirements

The fact that the computer has no direct contact with the process requires some consideration. First, data to be processed by the computer are usually generated by hand calculations and need preparation for computer use. Second, data and the processing program can be introduced at any time into the computer regardless of the actual condition of the process. Third, processed data need a method of storage for further utilisation

Manually collected data must be converted onto some medium which the computer is able to read, such as punched cards, punched paper tape or digital magnetic tape. Data on one of these media must be in the code form and in a given sequence specific to the computer and to the language used by the processor. The transcription of manual data onto the specific medium is usually performed through teletypes and the end product of the procedure can be loaded into the computer either by means of a card reader, or punched-tape, or magnetic-tape reader. If a time-sharing system with remote access terminal is used, data and computer program are collected on magnetic tape and processed according to the schedule specific to the time-sharing system. Processed data are listed in printout or on a medium compatible for computer retrieval. In all cases the identification of data is of the utmost importance. An excellent presentation of computer application in biochemical research along with the description of a computer protocol and the running instructions is given by Pizer *et al.* (1969).

Model Construction

The model formulated as a mathematical equation is the concise description of the particular biochemical-microbiological process for computer use.

General operating policies in model construction and model analysis are outlined in Fig. 3. At first, the problem is formulated as exactly as possible. Secondly, a model (or set of models) is constructed to express the problem in mathematical form.

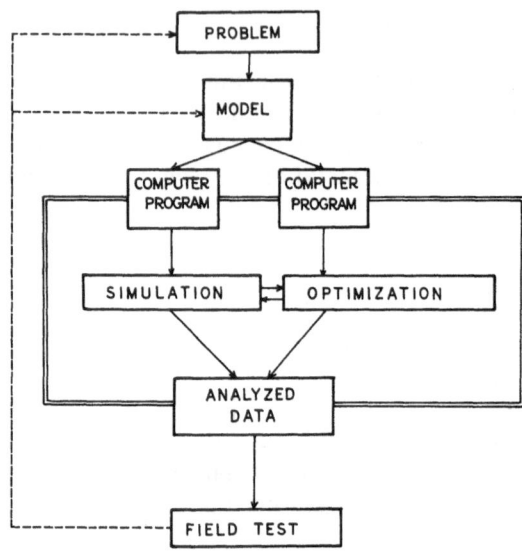

Fig. 3. Strategy of model construction and model analysis

Models dealing with microbial growth and cell metabolism belong to the symbolic model type (Ackhoff, 1962). These models use concise mathematical symbols to describe the way in which variables change and interact. Generally, models of cell growth and metabolic activity can be reduced to common forms of equations:

$$V_i = g(P_i) \tag{1}$$

$$P_i = f(C_i; U) \tag{2}$$

Eq. (2) is analogous to the general performance equation which describes the kinetic behavior of a system as a function of different environmental conditions:

$$\frac{dx_n}{dt} = f[x_n, b_n(C_n)]. \tag{3}$$

In a biological system the state variable (x_n) can be the growth, metabolite formation; the control variable (C_n) can be environmental factors such as temperature and nutrient concentration.

Table 4 presents the most frequently used basic mathematical equations derived from Eq.(1) and (2). These can serve as unstructured models of growth and product formation for process analysis.

It is necessary to emphasise, however, that the models of microbial cell growth and metabolism are oversimplified descriptions of the cellular

Table 4. Symbolic, unstructured models for microbiological process analysis

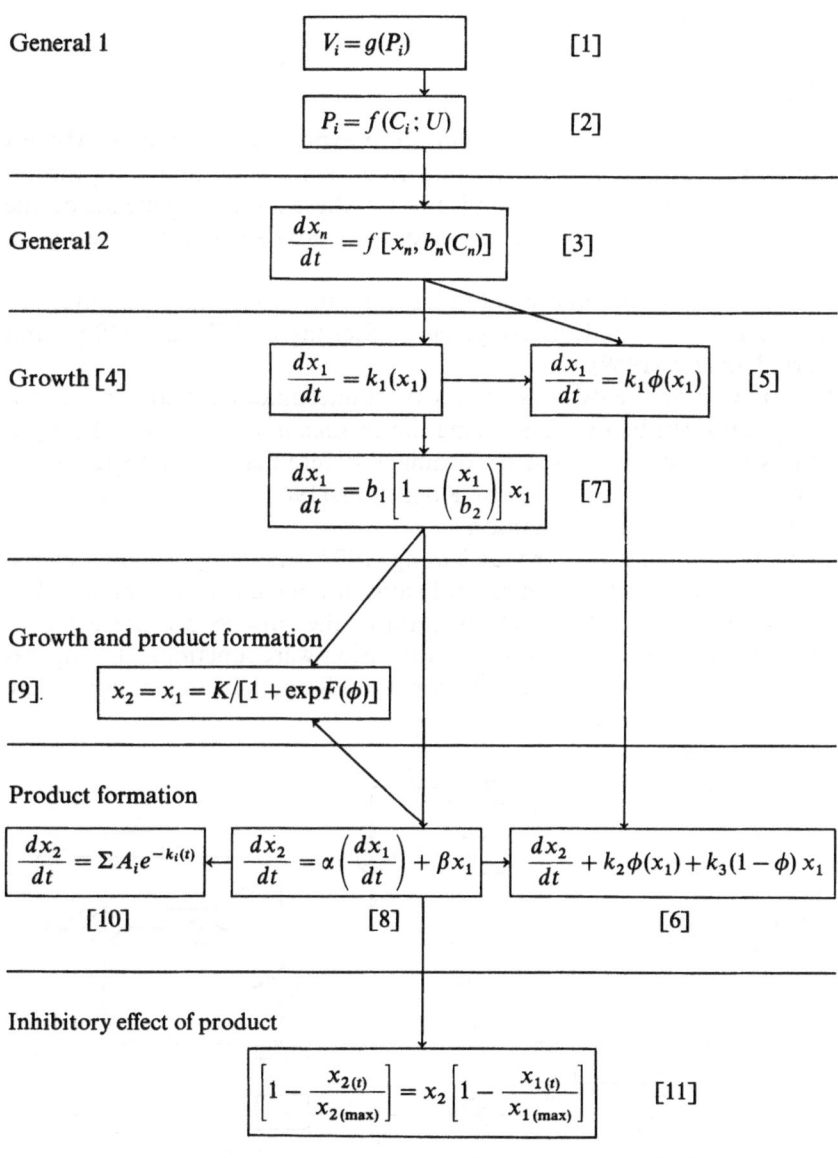

General 1

$$V_i = g(P_i) \quad [1]$$

$$P_i = f(C_i; U) \quad [2]$$

General 2

$$\frac{dx_n}{dt} = f[x_n, b_n(C_n)] \quad [3]$$

Growth [4]

$$\frac{dx_1}{dt} = k_1(x_1) \qquad \frac{dx_1}{dt} = k_1\phi(x_1) \quad [5]$$

$$\frac{dx_1}{dt} = b_1\left[1 - \left(\frac{x_1}{b_2}\right)\right]x_1 \quad [7]$$

Growth and product formation

[9].

$$x_2 = x_1 = K/[1 + \exp F(\phi)]$$

Product formation

$$\frac{dx_2}{dt} = \Sigma A_i e^{-k_i(t)} \qquad \frac{dx_2}{dt} = \alpha\left(\frac{dx_1}{dt}\right) + \beta x_1 \qquad \frac{dx_2}{dt} + k_2\phi(x_1) + k_3(1 - \phi)x_1$$

$$[10] \qquad\qquad [8] \qquad\qquad [6]$$

Inhibitory effect of product

$$\left[1 - \frac{x_{2(t)}}{x_{2(max)}}\right] = x_2\left[1 - \frac{x_{1(t)}}{x_{1(max)}}\right] \quad [11]$$

1. Ackhoff (1962). — 2. Ackhoff (1962). — 3. Generalized model. — 4. Monod (1949). — 5. Kono and Asai (1969). — 6. Kono and Asai (1969). — 7. Kendall (1949). — 8. Luedeking and Piret (1959). — 9. Edwards and Wilke (1968). — 10. Shu (1961). — 11. Friedman and Gaden (1970).

activity. Oversimplification is an inherent characteristic of any model of any process (Box and Hunter, 1962) and this is *par excellence* valid for living systems.

Model Analysis

The model itself is actually a procedure expressed in symbols. Models can be the subject of analysis using either simulation and/or optimisation or other methods. The application of these methods depends on the goal of process analysis. The choice of the analysis will define the sort of computer program.

General descriptions of useful fundamental mathematical manipulations related to model analysis are given by Sterling and Pollach (1968) and Carnahan *et al* (1969).

Information on the dynamic behavior of microbiological and biochemical systems can be obtained using *direct simulation* methods, the algorithms of which are based on ordinary differential and integral equations. This method was used by Koga *et al.* (1967) for computer simulation of gluconic acid production.

In his block diagram approach Young (1970) simulated the characteristics of interactions between several variables in a chemostat model. The general model and the block diagram of the process are presented in Fig. 4. Here growth and substrate utilization were functions of temperature, pH and time. The effects of growth on S uptake and *vice versa* were also simulated.

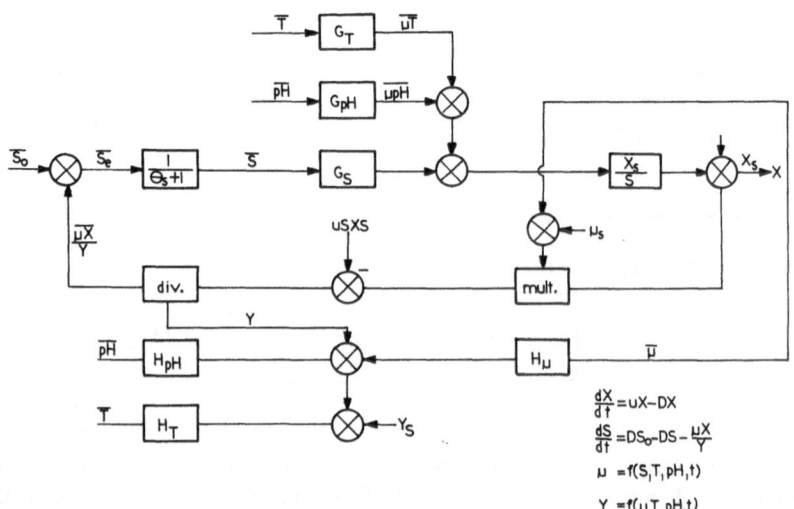

$$\frac{dX}{dt} = uX - DX$$

$$\frac{dS}{dt} = DS_0 - DS - \frac{uX}{Y}$$

$$\mu = f(S, T, pH, t)$$

$$Y = f(\mu, T, pH, t)$$

Fig. 4. Simulation of the dynamic behavior of a chemostat (Young, 1970)

A series of mathematical steps resulted in the transformation of these equations into the deviation equations, then into the Laplace domain. The deviation variables (denoted as superscript beta in the block diagram) were calculated by subtracting out the steady-state portion of the total variables from the total values. The G blocks relate deviation temperature, pH and S to the deviation growth rates. The H blocks relate deviation temperature, pH and growth to deviation yield coefficients. It is interesting to note that the $\overline{\mu X/Y}$ consumption term acts as negative feed-back control and thus represents a self-adjusting characteristic of the chemostat. Evaluation of the G and H transfer functions was made from dynamic experimental data obtained from a glucose-limited chemostat culture of $S.$ $cerevisiae.$

The situation is more complicated in the case of optimisation. The performance equations describing microbiological-biochemical processes always contain parameters such as velocity constants or equilibrium coefficients. Early investigations on microbial growth and metabolism indicated that the state variables are predominantly nonlinear functions of these parameters. In order to perform an optimisation procedure it is necessary to define at first the values of the parameters. Out of different methods used for parameter evaluation the non-linear regression techniques (Pennington 1970) were found useful and effective.

Algorithms of optimisation are based on differential calculus, extremum-finding procedures or direct search methods (Emhoff and Simsson, 1970).

The outline of an optimum-finding procedure is presented in Table 5. This procedure was used by Constantinides and his coworkers (1970a, 1970b) to define optimum temperature profiles for batch penicillin fermentations.

Flowchart Construction, Programming and Computation

The program construction for any computer operation first requires the formulation of a flowchart based on the mathematical equations of the model and of the analytical procedure. Since the flowchart depicts *what* is to be done more than *how* it is to be done it is a machine-independent graphical representation of the computational process.

Choice of an adequate computer and language for date processing are the next steps in the strategy. Application of large-scale computers has the advantage of the wide selection of mathematical software libraries and problem solving packages based on high-level, multi-purpose languages. This fact often makes possible the direct transcription of the flowchart steps into a set of program statements considering only the specific requirements of the language chosen.

Table 5. Outlines of an operation to find optimum value of a control variable

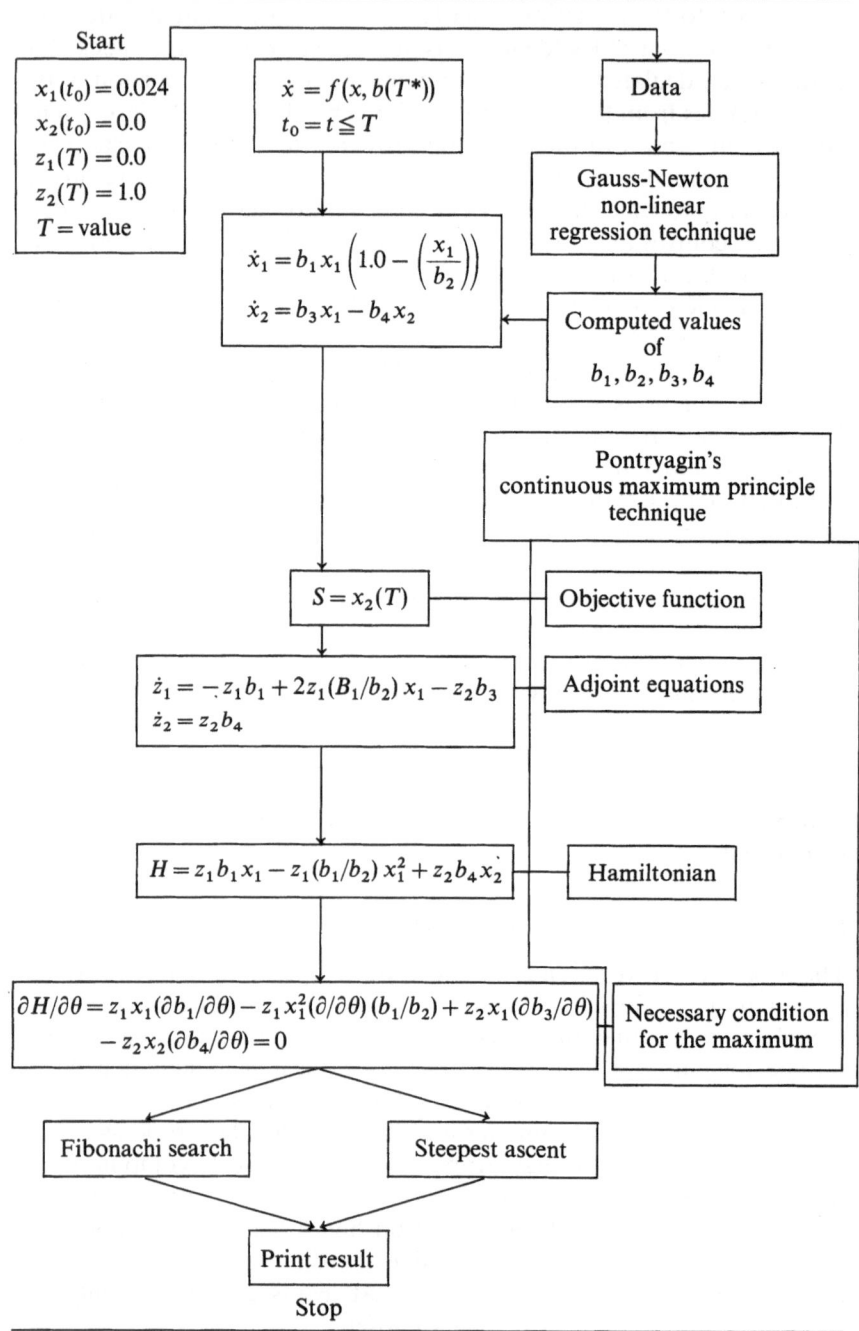

Field Test

Data obtained by means of any analysis are the subject of field test, which is, in this particular case, the comparison of the results with known and available experimental values of the fermentation processes. This fitting procedure can be done by computer using suitable algorithms and program languages (Hsieh et al., 1969; Naito et al., 1969; Edwards, 1970; Edwards et al., 1970).

On the basis of comparison the original model can be refined (e.g. new variables and parameters can be introduced) or reconstructed. The new model is then exposed again to another analysis. This iterative process of model construction and analysis is considered a powerful tool to improve knowledge of cell growth and metabolism.

b) On-line Operation

It is a generally accepted idea that the explicit goal of an on-line real-time operation of a computer is to perform more efficient control over the process.

Design of a highly automated chemical reactor or plant under computer control is based on that amount of knowledge and information which is available about the process. For instance, in the esterification of diacetyl ketogluconic acid with methanol and the subsequent, in situ rearrangement of the methyl ester to sodium ascorbate (Bacher and Kauffmann, 1970) the effects of control variables (in this particular case: the temperature, pressure, pH, ratio of reactants) were well known prior to industrial application. This knowledge can form the basis of a successful DDC operation performed by the computer.

One of the main characteristics of cell metabolism and microbial growth is the extreme complexity and the invariable possibility of the effect of environmental conditions on the process.

It is almost impossible to predict the impact of a recent change in a control variable on a later physiological status of the culture. Undesirable mutant formation, for instance, during batch and continuous cultivation of microorganisms has recently been considered by Aiba et al. (1965) as well as Kono and Asai (1969 b), but the effect of the environment on the nature of mutant formation has not been clarified. This complexity of microbiological processes needs more understanding prior to successful application of sophisticated control methods.

According to Humphrey (1971) in order to achieve meaningful control of fermentation processes it is necessary to

1. carry out fermentation research in a fully monitored environmental system,

2. correlate the observations with existing knowledge of cellular metabolic response,

3. refine the environmental control system according to the results of correlation. If necessary, start again with Step 1,

4. reproduce the obtained optimum environmental control conditions through continuous computer monitoring.

The on-line, real-time utilisation of the computer along with fermentation processes offers new scope for process evaluation: the instantaneous analysis of data available from the sensor systems during the fermentation. This has the following advantages, namely 1. the ability to keep track of the process from the earliest possible stage of the experiment and 2. to perform comparative analysis of the recent process status and former experimental data, and 3. the analyzed data will aid decision-making for process control.

The best place for the data-analysis-oriented, on-line utilisation of computers is the pilot-plant scale. The recent proliferation of small-scale computers (Theis and Hobbs, 1971) has placed this sort of data-processing system within the reach of pilot-plant scale application.

It seems reasonable both from the economical and practical points of view to carry out the experiments in pilot plants and *then* using large masses of well-analyzed experimental data, design the process control for industrial scale operation. This approach is a necessary evolutionary step simply because adequate environmental monitoring systems are yet non-existent (Humphrey, 1971). Literature data indicate the consideration of similar approaches to perform more suitable control of environmental conditions during fermentation processes (Yamashita *et al.*, (1969).

Physical and Organisational Requirements

Regardless of the orientation of the on-line computer utilisation the computer performs three basic functions, namely

1. Logging of process data,
2. Reduction of logged data, and
3. Process control.

Data logging is performed through the data acquisition system which has both hardware and software facilities. There is an interface (and multiplexer, if it is necessary) between the sensors and the computer. Choice of an adequate interface is of the utmost importance from both economic and computer function points of view. The software includes the computer program for sequential scanning of the sensor signals and for the procedure of data storage.

Data reduction is done by the data analysis system which is, in fact, a computer program based on an arrangement of suitable mathematical equations. The nature and complexity of the equations depend on the goal of data analysis. The data generated by the data analysis system are also the target of the data acquisition system from the data storage point of view. The analysed data ultimately can be utilised for process control.

Process control is performed on the basis of a computer program. Signals from the computer reach the actuators through an interface. In addition, the computer program contains instructions for utilisation of periferals such as display devices or teletypes. The format and production of hard copy of the collected data is also the subject of a program routine.

4. Computers and Programming Languages

a) Off-line Processes

Although analog computers offer some advantages with regard to their price and ability to cooperate with the user, the off-line data processings prevailingly performed by means of large-scale digital computers (cf. Table 1). These computers (e.g. the most frequently used IBM 360) are equipped with high-level compilers. This fact defines the possibilities of computer language usage.

The complexity of microbiological and biochemical processes along with engineering problems requires both problem and procedure-oriented computer languages. The multipurpose FORTRAN and its dialects (e.g. WATFOR) are found predominantly suitable to draw up schedules of instructions to deal with the problems of biochemical engineering. These languages give the user a relatively wide freedom of action allowing him to concentrate on the mathematical formulation of the problem instead of writing innumerable sets of instructions for different computer operations, such as floating-point mathematical operation or control of data transmission through input/output devices.

Cell growth and metabolism are time-dependent continuous processes. According to these the computer program languages with ability to solve differential-integral equations with continuous variables have been found to be particularly useful for process simulation purposes. The most frequently used CSMP (IBM, 1967), for instance, permits the digital computer to simulate an analog computer. This language gives the advantage of easy programming, is applicable to a wide range of problems in biochemical engineering and provides good error diagnostics. Fig. 5 represents a typical set of statements made for simulation of a batch fermentation model using CSMP language and an IBM 360/75 compu-

```
****CONTINUOUS SYSTEM MODELING PROGRAM****

***PROBLEM INPUT STATEMENTS***

*
          TITLE BACTERIAL GROWTH IN BATCH CULTURE
*
          INITIAL
*
          CONSTANT MUMAX=1.0
          CONSTANT KS=0.001
          CONSTANT Y=0.5
          INCON IX=0.05
          INCON IS=1.0
*
          DYNAMIC
*
          MU=MUMAX*S/(KS+S)
          DSDT=-MU*X/Y
          S=INTGRL(IS,DSDT)
          DXDT=MU*X
          X=INTGRL(IX,DXDT)
*
          TERMINAL
*
          TIMER FINTIM=10.0, OUTDEL=0.05, DELT=0.001
          FINISH S=0.0, X=0.6
          PRTPLOT X(S,MU)
          END
          STOP

OUTPUT VARIABLE SEQUENCE
MU      DSDT    S       DXDT    X
```

OUTPUTS	INPUTS	PARAMS	INTEGS +	MEM BLKS	FORTRAN	DATA CDS
9(500)	32(1400)	8(400)	2+ 0=	2(300)	6(600)	10

```
          ENDJOB
```

Fig. 5. Simulation program written in CSMP language

ter (Flynn and Corso, personal communication). This illustrates the sort of arrangement of statements as required by the CSMP language. The CSMP (or its equivalent: MIDAS) solves differential equations and integrations by routines based on the Runge-Kutta method (Pennington, 1970). Here the total execution time is determined by the step size of integration and the time required for the computation of the slowest part of the operation [in this case: computation of $f(x, y)$]. Erickson and coworkers (1970) describe erroneous results due to the improper step size of integration which may take place when parameter values are changed.

Edwards, investigating the inhibitory effect of high substrate concentration, utilized WATFOR to fit various functions to data (1970). The program consists of one main program and two subroutines: FUNCT and DFMFP. The main program monitored and controlled the input/ output functions and the number of iterations made by the subroutines. The FUNCT subroutine is used to calculate the value of the function to be minimised and the gradient of the function for any given choice of parameters. The function to be minimised is the error sum of squares calculated from the fit of any given equation to a set of data. The fittable parameters in the fitted equation are then varied by DFMFP to find the combination of values that minimise the error sums of squares, using a rapidly converging method described by Fletcher and Powell (1963). Edwards reported the necessity for modification of statements in FUNCT subroutine if a different mathematical function is to be fitted to the data. The program was found to be quick to converge to the best fit for three-parameter functions, but relatively slow when a four-parameter function was fitted to a set of experimental data. The whole program was found more stable than some nonlinear least squares programs (Edwards and Wilke, 1968).

With increase of the number of variables and constants (Heinmets, 1969; Megee et al., 1970; Tanner, 1970) the complexity of the investigated model system increases. In other cases the user needs further specific evaluations of the data processed by the computer using a given language. In these cases there is a need for introduction of other, compatible languages or routines into the program. This was the case, for instance, when CSMP was used the compiler of which makes possible the joint utilisation of FORTRAN IV (Nyiri et al., to be published). This process results in extremly flexible program structures producing "hybridisation" of analog and digital computation techniques using digital computer and suitable program for this purpose.

There are examples in the literature describing specific computer programs as well as evaluating the applicability of some mathematical methods and softwares used in biomedical science (Rhyne et al., 1970) and biochemistry (DeLand, 1967; Sheppard, 1969; Garfinkel et al., 1970). Some of them can be used in biochemical engineering too.

b) On-line Processes

Considerations for choice of a particular computer to perform the on-line logging, data-reduction and control functions in fermentation processes include (Nyiri, 1971a):

1. Suitable word capacity which assures the handling requirements of the whole system,

2. An expandable memory capacity which accommodates future needs with specific regard to decision control loops,

3. A large selection of peripheral equipment, interface boards and software library, and

4. Adequate field service ability.

Depending on the number and type of available computers and their memory capacity there are several variations for time-sharing processes. An economical alternative is data acquisition by small-scale computer (using 4 K—8 K core memory for program execution) followed by data transmission by the small-scale computer to a central computer utility. Data analysis takes place in the large-scale computer. In any case, an on-line, small-scale computing system is absolutely necessary if powerful techniques for data acquisition are to be employed (White and Hazbun, 1970). Because of the high installation costs, utilisation of large-scale computers for process control is economical if several fermentors (i.e. economically sufficient amount of control loops) are monitored (Cotter, 1969).

The on-line utilisation of computers along with fermentation processes requires multipurpose languages which assure program constructions for computer operation related both to 1. process-computer, and 2. computer-operator interactions. Any kind of languages suitable to the computer in question can be useful for the first purpose. For instance, in the case of joint use of a small-scale computer with a pilot-plant scale fermenter the computer program was written in assembly language (Nyiri, 1971a). Nevertheless, in the case of small-scale computers the actual core memory size and the compiler availability are the factors which define the type of language used.

The interactions between the computer and the operator are performed through dialogues. There are some criteria which ought to be considered in constructing dialogues for fermentation process application, namely:

1. Dialogue must be as short as possible, assuring quick actions,

2. Text of dialogue must contain mnemonic and symbolic elements easy to learn and handle through the keyboard,

3. Although the text is highly specific, the computer must be programmed to ignore minor typing errors (such as double spacing instead of single in the text).

A typical dialogue between the operator and the computer is presented in Fig. 6. The text was written for a PDP-11/20 small-scale computer connected to a pilot-plant scale fermenter. Here the operator instructs the computer to change the pH of the culture liquid at a given time during the fermentation. The dialogue consists of statements each written in one line. First the operator identifies himself (OP) which causes termination of execution of the job in progress. The computer prints */ characters indicating the clearing of channels for the input device in order to accept further inputs. The operator then defines the sort of operation (in this case CHANGE OF A CONTROL VARIABLE = CHNG). The computer expresses its ability to select the subroutines for adequate operation (*?). The operator then defines the sort of variable (PH) then the time of alteration (T00000), the lower and higher setpoints (LSP and HSP, respectively), then terminates the dialogue (OP←). The computer finally acknowledges the receipt of the message (*OK). A practiced operator can accomplish this instruction in less than 30 sec.

```
OP
*/
CHNG
*?
PH
T4200  LSP  6.8;  HSP  7.0
T4800  LSP  7.1;  HSP  7.2
OP←
*OK
```

Fig. 6. Dialogue between the operator and the computer

5. Data-Analysis-Oriented Use of the Computer

In this chapter a strategy of computer program construction is depicted. This was used during the design of a data-analysis-oriented, on-line utilisation of a small-scale computer along with a pilot-plant scale fermenter. The system is going to match the requirements of application of computers for extended, detailed, real-time process analysis.

a) Layout of Facilities

Table 6 and Fig. 7 represents the major elements of the fermenter-instrument-computer-logic complex with specific regard to the relationship between the fermenter and the computer.
Two sensor systems are distinguished depending on the sort of information available through them (Table 6):
1. Physical sensors, which detect the status of controllable process variables,
2. "Gateway" sensors (Humphrey, 1971) which detect the status directly or indirectly related to state variables.
Combination of values of the two sensor systems using adequate mathematical models gives further information on the dynamic behavior of the system.
As is seen in Fig. 7, sensor signals from the fermenter are amplified (AMP) and displayed on digital panel meters (DPM). A noteworthy feature of the hardware design is its modular construction (Harmes, 1971). Instead of using conventional A/D converter the individual DPM

units are utilized to perform the conversion of analog signals to binary coded decimal (BCD) signals and to transmit them to the interface. The construction of the computer-fermentor (output) interface assures a computer access to the fermentor *via* limit switches or set points on the

Table 6. Sensor systems for an on-line computer utilisation with fermenter

Physical sensors	"Gateway" sensors
Thermometer	pH electrode
Pressure transducer	Redox potential electrode
Flow meters	Dissolved oxygen electrode
a) gas flow meter	Oxygen gas analyzer
b) liquid flow meter	Carbon dioxide gas analyzer
Speedometer	Automatic analyzers
Wattmeter	Turbidometer
Dynamometer	
Load cells	
Foam detector	

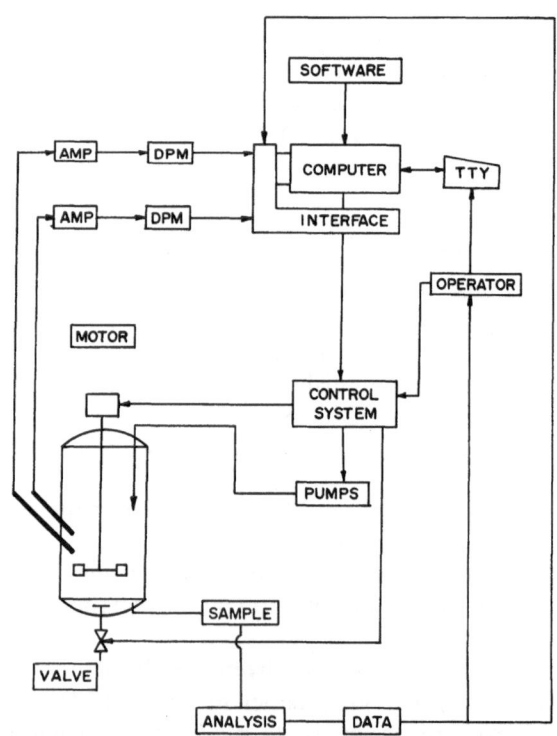

Fig. 7. General layout of pilot-plant scale fermenter

controllers. These individual, closed-loop type controllers actuate pumps and valves on a real-time basis to alter the control variables. This concept obviously requires a highly instrumented fermenter with perfect performance of operation from instrumentational points of view (Fig. 8).

Fig. 8. A highly instrumented pilot-plant scale fermenter (Department of Chemical Engineering, Pennsylvania State University, Philadelphia, Pa.)

b) Computer and Software Capabilities

Fig. 9 indicates the arrangement of the computer with respect to its major functions. It is noted that the data-logging and data-reduction systems operate independently from the process control system.

The recently developed PDP-11/20 computer is equipped with an assembler based on instruction sets specific to this device. This choice defined the language used for the programming.

c) Computer Program

A general flowchart of the whole program structure is presented in Fig. 10.

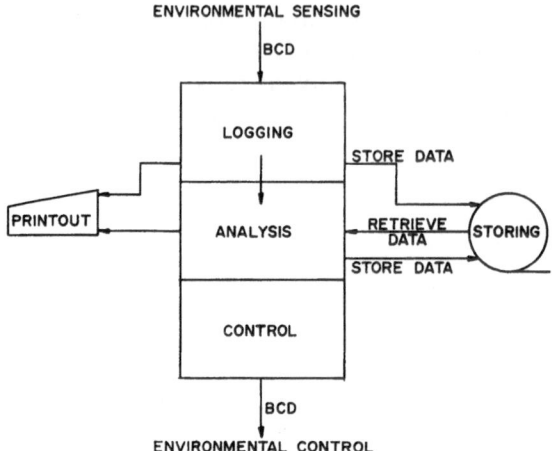

Fig. 9. Computer and its major functions

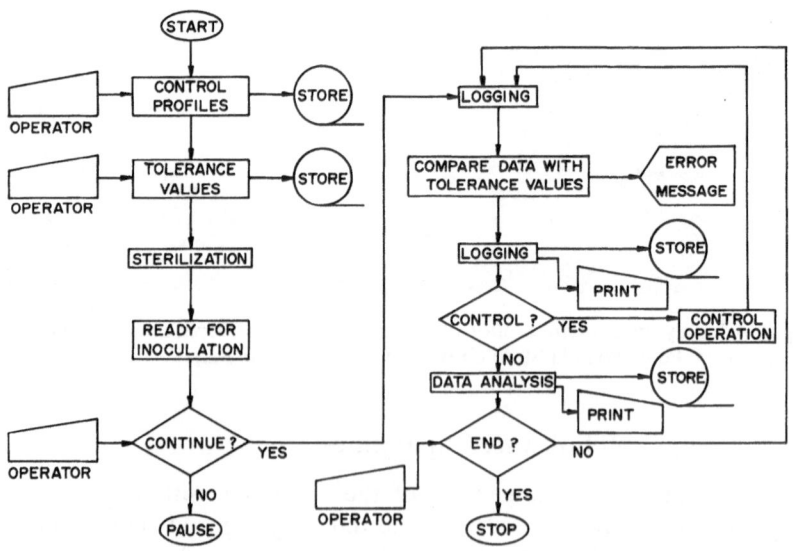

Fig. 10. Essentials of computer program for data-analysis-oriented operation

The program package consists of two major portions, namely
1. The *preparatory phase*, where two basic functions are performed:
a) calculation of tolerance values of the process variables which are
introduced through the operator's console, and b) sterilization. In this
preparatory phase the program sequence has no major program loops

and its function can be suspended when the fermentor is ready for inoculation.

2. The *executive phase* contains the program routines for data logging, data analysis and process control.

Data Logging

Sensors' data are scanned sequentially in time and are identified with reference to the real time of the fermentation process. Two program sequences were constructed based on efforts to save computer time for data analysis instead of using it for printout function. Accordingly, sensors' data are scanned more frequently than they are printed on teletype. In the first scanning cycle the actual sensor values are compared to the preset and tolerance values, and deviation from these values results in an error message printout. The second data-logging program sequence includes the commands for scanning of sensors' data in a given time followed by storage and printout of data.

Data Analysis

One of the main objectives of computer utilisation is to increase the number of observations on the particular process providing thus insight into the physico-chemical, physiological, and biochemical conditions of the process which would otherwise be either unrecognised or could be determined only after the actual condition had passed.

Requirements of an adequate structure for a data analysis program for on-line utilisation are (Nyiri, 1971 b):

1. Program consists of suitable algorithms,
2. Program construction is based on the principle of directed graph method,
3. Program assures the access of the central processor to any data obtained previously from any source for multivariation,
4. Execution of any part of the program can be temporarily suspended by the operator without influencing the execution of other parts of the program.

According to these requirements algorithms to calculate physical, physico-chemical, physiological and biochemical characteristics of the system (Table 7) were arranged in sequences resembling a graph tree form (Pennington, 1970). Here each node of the graph tree representation covers a series of mathematical operations leading to a value which appears at the exit line of the node and serves as a basis for the next step operation. Utilisation of the concept of the directed graph method made the whole program structure open-ended assuring introduction of new algorithms. An executive program investigates the availability of all sensor data,

Table 7. List of algorithms for data analysis

Physical	Physico-Chemical	Physiological	Biochemical
Power input[a]	Apparent viscosity	Oxygen uptake	Carbon balance
Rate of addition of ingredients	Power number	Carbon dioxide release	Organic energy yield
Volume of fermentation liquid	Reynolds number Flow characteristics Power characteristics Volumetric absorption coefficient	Respiratory quotient	Cell mass[b]

[a] Torque, Watt.
[b] On basis of — acid/base titration and/or
— CO_2 production,
— turbidity measurement.

parameters and constants necessary to carry out the particular analysis as well as checking the existence of a prohibitor signal which also can prevent the execution. In the case of halting of execution of a program sequence, the possibility of performing the next analysis step is investigated. In a data multivariation program, data of the former fermentations are retrieved in order to compare them with the recent conditions. Three main areas of fermentation process are mentioned here; all of them form the basis of the data analysis program. These are:
1. Rheological conditions of the culture liquid,
2. Physiological status of the culture, and
3. Biochemical characteristics of the cells.
Fig. 11 depicts the arrangement of the algorithms for real-time analysis of variables related to the rheological conditions. Logged data are reduced to give instantaneous information on the relationship between the actual rheological and mass-transfer conditions in the particular culture liquid. This information on rheological conditions can also be utilised for process control and scale-up procedures.
The application of reliable oxygen and carbon dioxide gas analysis devices made it possible to construct programs for detailed, real-time study of oxygen uptake and carbon dioxide evolution of microbial cultures. In this respect the fermentor operates as a differential respirometer. Fig. 12 shows the basic idea of the study of some physiological characteristics of the cells with the assistance of the computer.
Oxygen uptake and carbon dioxide evolution by the culture can be related to obtain information on the actual respiratory activity of the culture. The respiratory quotient and data on carbohydrate metabolism and mass-transfer coefficient can be interrelated to give decisions for

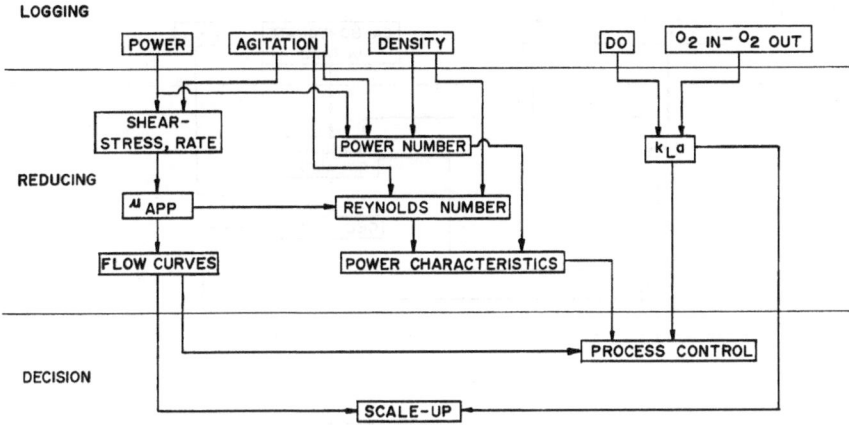

Fig. 11. On-line determination of some physico-chemical characteristics of the fermentation

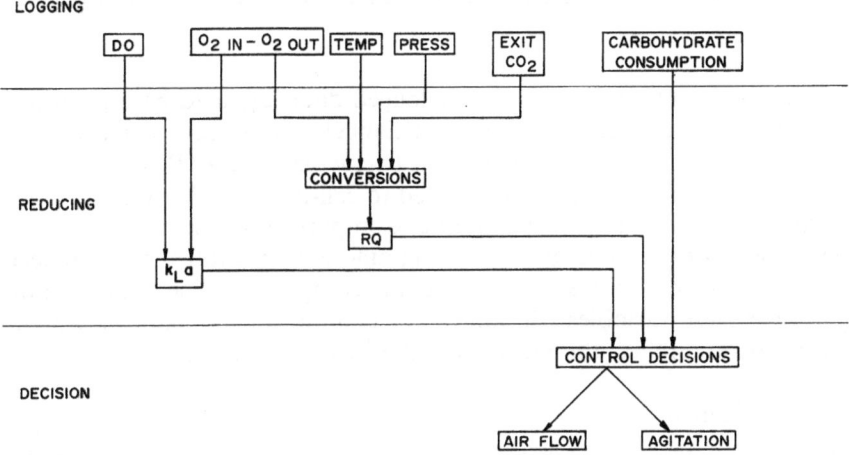

Fig. 12. On-line determination of some physiological characteristics of the fermentation

process control. For example, the effect of carbon dioxide on the metabolic activity of the cells (Zajic, 1964; Nyiri and Lengyel, 1965; Lengyel and Nyiri, 1966; Nyiri, 1967; Wimpenny, 1969) can be considered to find the proper balance in controlling the airflow and agitation speed.

Information on respiratory activity and carbohydrate metabolism is valuable in understanding the metabolic pathways of particular product formation. As is seen in Fig. 13, computations can be performed to determine the changes in carbon balance, to obtain estimates of cell-mass

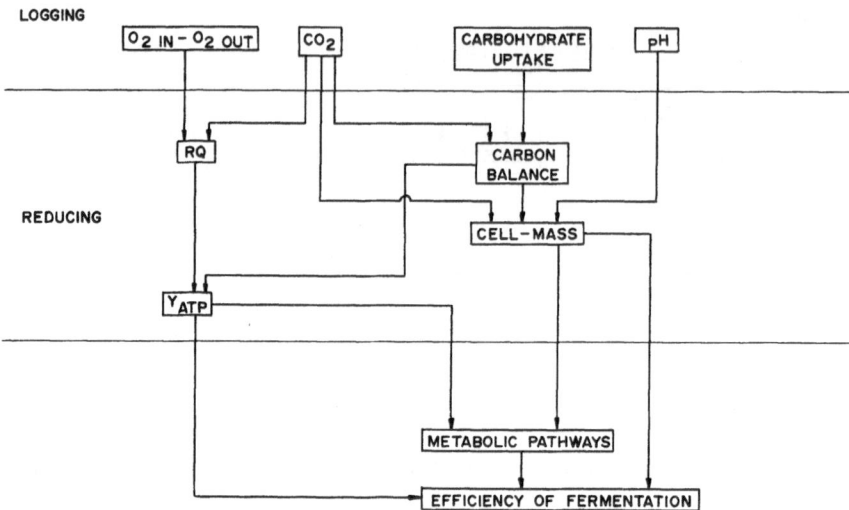

Fig. 13. On-line determination of some biochemical characteristics of the fermentation

production using different suitable logged data (e.g. acid/base titration, carbon dioxide evolution, carbon balance). Knowing the pattern of cleavage of carbohydrate it is possible to calculate the ATP yield (Aiba *et al.*, 1965) which, in turn, can be related to cell-mass production or to the formation of a particular metabolite. Comparison of the computed cell yields with organic energy yield, using different substrates, might reflect the mechanism of breakdown of organic compounds. The relationships between these variables will aid the elucidation of the main metabolic pathways and reflect the efficiency of the fermentation process.

Process Control

As is seen in Figs. 7 and 9 the computer does not participate directly in control. It has access to the set points or limit switches of individual controllers which perform the maintenance of control variable values at the desired levels without the assistance of the computer. The control variables and environmental systems are presented in Fig. 14.
The program for the computer to alter the set-point or limit-switch positions on the individual controllers has real-time instructions to establish a time-based control profile of control variables. Values of the profile are introduced into the computer through the teletype by the operator prior to the fermentation. The operator can alter the control variables values at will, any time during the operation (cf. Fig. 6). Also the control program can be readily changed on the basis of continuously

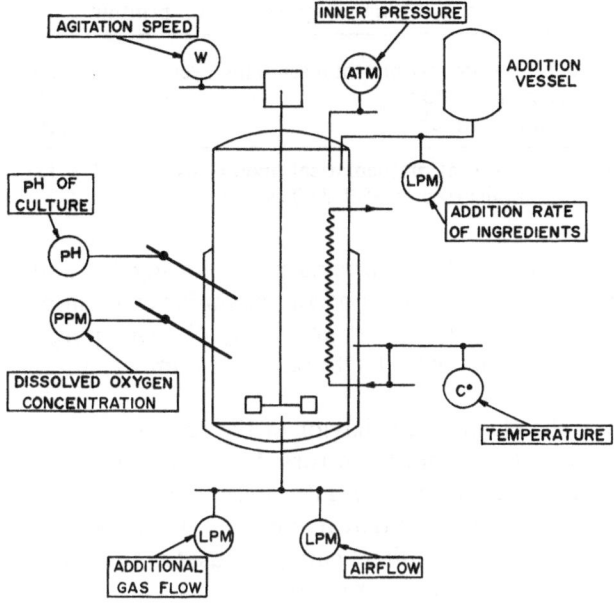

Fig. 14. Controlled process variables

accumulating experimental data. The control system using control profiles fulfils the requirements of stage-wise operations. In addition to this,
the fact that the operator can override the previously described control
profile gives the opportunity to perform on-line, real-time investigations
on the effect of environmental factors on the metabolic activity of the
cells.

d) Functional Considerations

The program was designed to be as general as possible, allowing the user
to "tailor" the program sequence to the requirements of the particular
experiment.
As usual, the execution time of the total program was specified for basic
mathematical operations. Table 8 indicates the processing times in one
scanning cycle using floating-point mathematical operations with PDP-
11/20. The basic mathematical operations take about 0.1 sec. The printout of characters which cover the record of all data obtained through
data logging and data reduction takes 120—140 sec. Batch fermentation
processes and critical changes during the fermentation are relatively
slow compared to a chemical reaction. By choosing a 5-min. interval for
the scanning and comparison of scanned values with tolerance values

Table 8. Processing times during computation and printout

Computation time within total scanning cycle[a]	600000 μsec
Characters in printout	1400 ± 10%
Total printout time[b]	120—140 sec

[a] Based on floating-point mathematical operations with PDP-11/20 computer.
[b] Assuming the application of ASR-33 Teletype.

and a 15-min. interval for scanning values for logging and data analysis plus printout there is ample time and room for expansion of the computations. Moreover, there is the possibility of developing a time-sharing system allowing the computer to service a number of fermentors at the same time.

Analysis of the electronic behavior of the sensors is of utmost importance to avoid unrecognisable errors during data sampling. The signal, which represents a time series is introduced to a digital panel meter. One advantage of this device is that it can filter the masking noises and can produce BCD signals for direct computer usage. Accordingly algorithms for nonlinear filtering are not required.

Environmental conditions can effect the response of sensors. Early experiments indicate, for instance, that it is necessary to build a temperature compensation system that electrically matches the temperature behavior of galvanic type oxygen probes equipped with diffusion membranes.

A specific set of experiments is required to define the delays between the actuation of an effect and the system's response. Although any response in the living system related to the environmental effect is specific, responses on physiological and biochemical levels have shorter time delays than responses manifested in growth and mutation. Specific algorithms based on off-line simulation experiments are needed when environmental effect-metabolic response relationships of the cells are investigated. These algorithms deal with the time delay of the specific responses.

e) Validity Test of Experimental Results

With the application of the computer the certainty of conclusions requires the methodical testing of the accuracy of the individual experimental data. The program package contains specific program sequences able to filter random electric noises (if any), to compute the mean values, the variance and the possible standard deviation for each measured and computed variable. Continuous experimental data in the fermentation industry are generally of a stochastic character. Analysis of variance of data was found to be a powerful technique to evaluate the data of multidimensional experiments. A specific program is required to define the goodness to fit the data to a curve based on a hypothetical mathematical model, which represents the working hypothesis of the experiment. The program also consists of statements of the suitable truncation of the digits.

These computations occupy significant computer memory and time. However, these operations are considered to be fundamental for adequate control decisions.

6. Process-Control-Oriented Use of Computers

Replacement of conventional analog controllers by a digital computer results in a computer-executed closed-loop, direct digital control of process variables. The DDC itself is the same as is performed in many cases in the petroleum and chemical industries. An oversimplified flowchart of the DDC performance on one control variable is presented in Fig. 15.

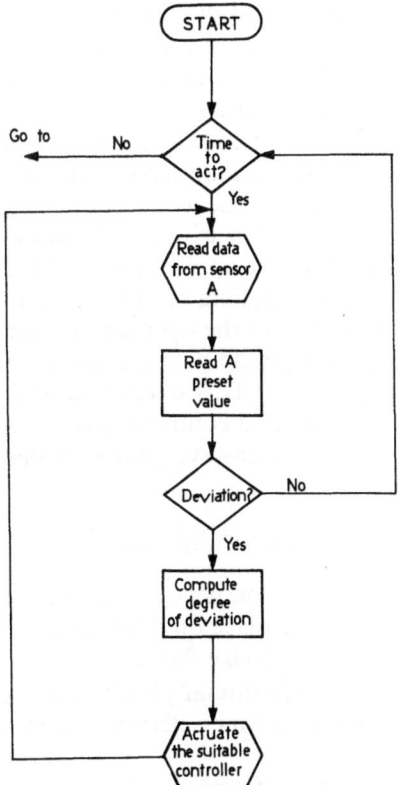

Fig. 15. Essentials of a direct digital control system

This indicates that data acquisition and data analysis systems also existing here. The functions of the data analysis system are 1. to compute the
degree of deviation between the preset and the actual values of the
process variable and 2. define the degree of response to actuate the
control device. Algorithms depend on the type of controller and their
control function. According to this, the computer is an organic part of
the control loop. The individual control loops are working in a non-
interactive way.

a) Evaluation of the Use of DDC in Fermentation Processes

DDC application in the fermentation industry has been found effective
and economical in batch fermentation processes when a relatively large
number of fermenters and a relatively small number of control variables
are under control. Sequence control could also be performed as a part of
the computer program.

During data acquisition the sensor signals have been measured by trans-
ducers, amplified, multiplexed and discretised. Background noises were
reported (Yamashita *et al.*, 1969), but the 99.95% availability during the
operation indicates that thought and care was taken to remove the
process noises by appropriate filtering.

In the start-up process DDC provided sufficient performance to remove
excessive overshoot and produce smooth rising-up action. Since no drifts
were found in control parameters the whole control operation was more
consistent compared to the action of conventional analog controllers.
The description indicates that the application of DDC made it possible
to reduce human errors in controlling several process variables. A suitable interface makes it possible for the operator to alter the preset values
of process variables (e.g. change the temperature during the process) or
the constants of the algorithms. This offers a certain flexibility in the
control performance. Coordinated control of process variables, depending on the status of metabolism, however, have not been reported as yet.

7. Future Trends of Computer Application

As is known, the effects of environment are reflected in changes of cell
function. The changes can take place on different levels such as physical
[e.g. alteration of cell shape and size (Wyatt, 1970)], physiological (e.g.
change in extracellular and intracellular pH), biochemical (e.g. change in
rate of enzyme synthesis and enzyme activity), and molecular biological
level (e.g. mutation).

Some of these changes can be detected using physical and "gateway"
sensors as well as by making physical and chemical analyses and calcula-

tions. Thus certain information is available on the chemical composition of extracellular space from which the metabolic activity of cells can be deduced. Less is known about the intracellular conditions, their interactions and their response to the influence of environmental changes.

The need of refined and more frequent measurement of the most important state variables such as substrate utilisation and metabolite formation led to the extended application of Autoanalyzers and the different enzyme-activity measuring devices as well as instruments with the ability to measure cell density directly (Moss and Bush, 1967). In recent years considerable engineering effort has been made to solve the problem of automatic sampling (Harmes, 1971) and filtering the culture broth sample which is considered to be the last "gap" in the complete automatic analytical follow-up of some characteristics of microbiological processes in liquid cultures.

Measurements of NAD—NADH levels in whole fermentation broth using a fluorometric method (Harrison and Chance, 1970), and the luciferase analysis of ATP (Hastings, 1968) give information on the actual catabolic and anabolic activities of the cell.

The total number of sensors is, however, limited. It is necessary therefore to maximise the information available through multivariation of the existing physical and gateway sensors.

Mathematical models on intermediary metabolism (Garfinkel *et al.*, 1970; Tanner, 1970; Megee *et al.*, 1970), on the dynamic behavior of growth and product formation (Kono and Asai, 1969a, 1969b; Young *et al.*, 1970), on the effect of environmental factors on growth and metabolism (Constantinides *et al.*, 1970a, 1970b) as well as on the predictability of the fate of the fermentation process (Gyllenberg *et al.*, 1969) have been tried successfully. These models along with data reduction algorithms can be used to program computers for detailed process analysis. In this case reduced process data can be compared by the computer to (average) data of previous experiments and with the values obtained by simulation, optimisation and prediction procedures. According to this the data-analysis-oriented computer operation on pilot-plant scale can serve as a tool of extended investigations on the very complex problem of environmental effect—metabolic response. Algorithms obtained by this means will serve, finally, as the basis of a control system which really matches the complexity of the microbiological-biochemical processes: *interactive control.*

Interactive control is performed by a multivariable control system, where the values of control variables are changed in an interactive way to achieve a desirable effect on the biological process. The decision to perform interactive control is made by the computer's central processor

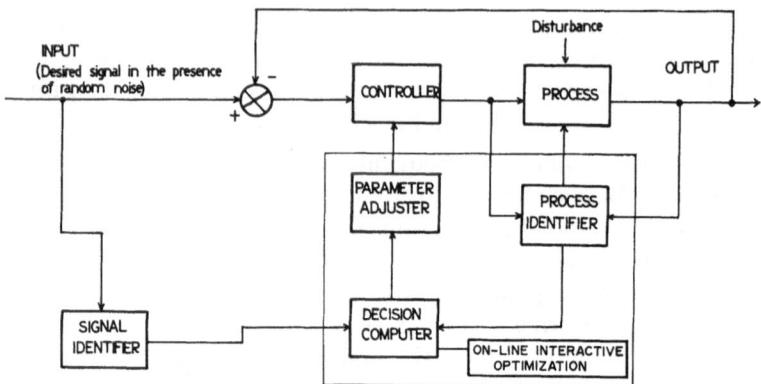

Fig. 16. Interactive control system

using a software based on algorithms which define the necessity, direction and degree of interaction among the control variables. Fig. 16 illustrates the general arrangement of elements in an interactive control system. Interactive control needs specific algorithms which can identify and optimise the process dynamics. According to this two necessary steps must be considered, namely:

1. Establishment of the process-identification method,
2. Development of on-line optimisation processes.

In the case of fermentation processes the identification of the status of the process can be made by direct sensor readings and/or indirect chemical analyses of samples. Future development and utilization of some essential "gateway" sensors (e.g. turbidometer, NAD—NADH, ATP-level measuring devices) can enhance the information on the status of intermediary metabolism. Recently it was demonstrated that computerised on-line data-logging and data-analysis system, based on the individual sensor readings, can increase the number of observations and can provide real-time information on the process. A proper description of the status of the process from the biological point of view is considered as the first essential step in the direction of adequate process dynamics identification.

Values describing the status of the process can be fitted to values obtained through simulation of the process in question. By this means the future (dynamic) trends of the process can be identified.

There are several difficulties from the technical point of view in the process dynamics identification. These are generally related to 1. the measuring equipment with respect to the accuracy and the noise level of the signals and 2. determination of the time interval of measurements to define the effect of the control variable on the state variable.

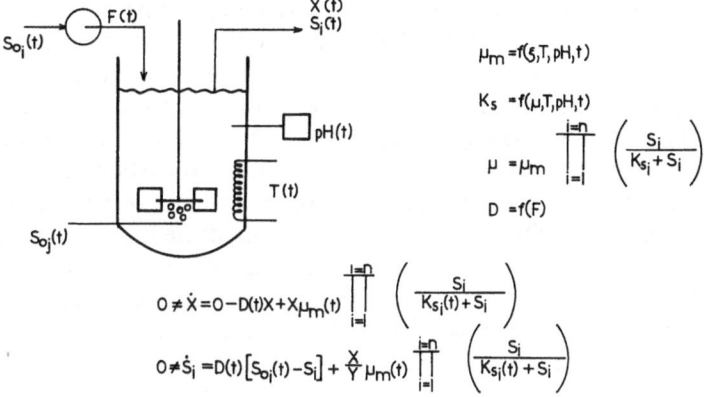

Fig. 17. A complex chemostat

Out of the fundamental difficulties in the construction of suitable process identification algorithms one is specific to the microbiological-biochemical processes, namely the determination of the characteristics of metabolic responses from a mathematical point of view. It is a generally accepted idea that the processes related to living systems are nonlinear and have random variables. The mathematical handling of this sort of system is extremely difficult even with computer assistance.

Generally, there are two major situations to be considered in fermentation processes:

1. Interactions among various control and state variables,
2. Interactions among the elements of cell machinery.

Fig. 17 presents a model of a complex chemostat, where the growth rate (μ) is theoretically affected by limiting substrate concentration, temperature and pH (Humphrey, 1971). Equations presented in the figure indicate that there is a highly unstable condition existing throughout the growth, because of the interactions among the control and state variables.

This case is considered as a nonlinear, interactive system. However, the control variables have Gaussian characteristics; furthermore, there is a possibility of finding finite-dimensional vectors for each variables (μ, S, T, pH) which, when specified, can be used in a differential equation that governs the transition from one instant to another. In this case equations which describe the interactions can be amenable to deterministic treatment. From the computational point of view this situation is more convenient. Here a trial using e.g. Young's block diagram approach (1970) along with computerised pilot-plant scale operations and trial-and-error method of process analysis (done by computer) could reveal the true characteristics of the process.

Metabolism can change during the life cycle depending on various conditions within the cell. It has been observed, for instance, that young *Streptomyces rimosus* hyphae which are not able to produce oxytetracyclin use the Embden-Meyerhof pathway for glucose catabolism. This pathway is sensitive to oxytetracycline. With the start of oxytetracycline production *Streptomyces rimosus* hyphae shift their carbohydrate metabolism to the oxytetracycline-insensitive hexosemonophosphate shunt (Nyiri *et al.*, 1963). These two stages of carbohydrate metabolism require different control profiles for temperature, aeration, agitation and pH. This case is complex from the mathematical treatment point of view. Here the conditional probability of the change in metabolic pathway is virtually independent of the past condition and depends on the present state of the process. So there exists a probability-density function of a finite-dimensional vector the definition of which is complicated. There is a lack of immediate indication of change of metabolism which makes the recognition of the status of the process difficult. Moreover, there is a need for adjustment of the environmental conditions during the changes in the metabolism which is the prerequisite of a high antibiotic yield.

The behavior of fermentation processes has been the target of innumerable investigations from both the theoretical and practical points of view. Studies have revealed that 1. there are stages during the process, 2. each stage of the process can be characterised by specific metabolic activities, 3. analysis of dynamic behavior of growth and metabolic activity in each stage requires different mathematical treatment, and 4. different environmental conditions are required for the different stages.

Attempts to fit theoretical and experimental data for different fermentation processes have been found successful in many cases (e.g. Koga *et al.*, 1967; Edwards and Wilke, 1968; Gyllengerg *et al.*, 1969; Kono and Asai, 1969a, 1969b; Young, 1970). These facts indicate that there are possibilities of constructing algorithms for process-dynamic identification by which the individual stages of the process could be identified with acceptable accuracy. Hence, the optimum environmental conditions for the given stage can be defined. The environmental and molecular biological conditions which induce the transition from one stage to the next one, however, seem to be worthy of further investigations. Here mostly the interactions among the enzyme systems on the molecular level and the effects of environmental factors (Koga *et al.*, 1969) are of importance.

Adequate process identification is essential to the choice of optimisation method. Responses of Gaussian character could be treated by methods based on statistical decision theory. In the case of nonlinear systems, multidimensional joint statistics (e.g. Bernoulli trial) might be a useful approach (Sawaragi *et al.*, 1967).

Another approach to the optimisation of fermentation processes is the *pivot-finding strategy*. Here it is assumed, that there is a mechanism in the process which is the pivot of the whole process and, although this mechanism is governed by several control variables, there is a pivotal control variable on which the mechanism depends. For instance, respiration is considered to be a pivot biological state variable in the aerobic cell metabolism. In some cases temperature was considered (and found) to be the pivotal control variable (Ho and Humphrey, 1970; Constantinides *et al.*, 1970a, 1970b). Using this strategy after the proper process identification, the actual pivot mechanism and its control with a pivotal control variable can be the subject of the optimisation procedure. This strategy, however, requires extended investigations on the tolerance latitudes of state variables to the effects of individual control variables.

The next problem is the evaluation of environmental systems from the viewpoint of their utilisability for interactive control. Owing to the specificity of living systems only those variables which have the following characteristics could be considered as powerful tools:

1. there is a measurable (or computable) response to the effect,
2. the response is proportional to the effect,
3. there is a maximum or minimum in the effect response curve,
4. there is no irreversible damage in the system when the effect is actuated.

According to this, generally, the following systems are potentially usable for an interactive control (assuming aerobic culture with the dissolved oxygen concentration above its critical value):

a) Temperature,
b) pH,
c) Concentration and ratio of nutrients and precursors,
d) Dissolved CO_2 (HCO_3^-).

The effects of these control variables are manifested through the growth and metabolism of the cells. Among the different metabolic activities the respiratory activity of the cells is of primary importance.

Recently a computer program was developed for detailed analysis of the available physical and "gateway" sensor signals (Nyiri, 1971a, 1971b). This results in instantaneous information on the $k_L a$ value, the respiratory quotient, the direction and the slope of the change of these values, the rheological characteristics of the culture broth, on the basis of agitation speed, power input, rate of airflow, the actual dissolved oxygen concentration and the rate of carbon dioxide release. These data provide parameters for the evaluation of the effects of control variables on respiration and the relationship between respiratory activity and certain primary and secondary metabolic activities.

The biological interactive control system is related to some extent to the adaptive control (Li, 1960). The application of adaptive-type control systems is in its infancy even in less complicated cases than the biochemical processes (Cotter, 1969; Bristol, 1970). Nevertheless, it is not unrealistic to consider the construction and utilisation of algorithms, made on the basis of recent knowledge of fermentation processes, for interactive control purpose, at first for relatively simple, then more complicated fermentations. Fig. 18 depicts an oversimplified flowchart which includes the above-mentioned necessary steps of process identification, optimisation and process control. This may be useful in the development of interactive control systems in the case of fermentation processes.

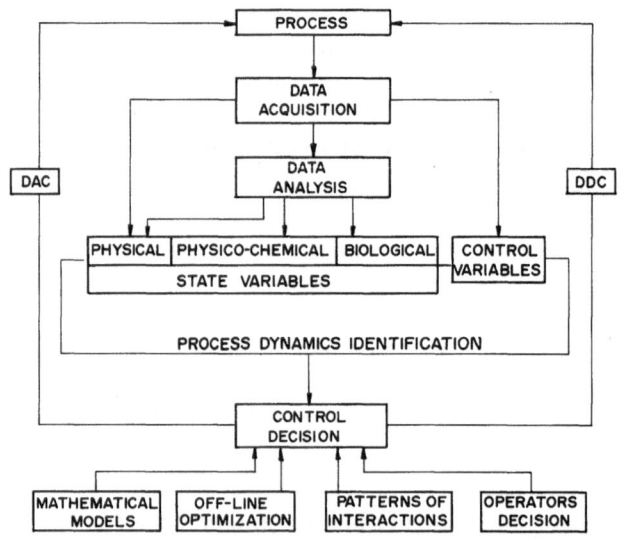

Fig. 18. Essentials for development of a biological interactive control system

8. Conclusions

Computers are useful in handling and interpreting large masses of data. Since the problems of biochemical engineering involve living systems of complex nature the utilisation of computers results in better understanding and manipulation of fermentation (microbiological-biochemical) processes.
Application of computers in the fermentation field is just beginning. Their utilisation, no doubt, will accelerate with the increased availability and reduced costs of suitable small-scale, general-purpose computers.

As a consequence of computer application, specific methods are necessary to collect, analyse, store and exchange experimental data. The choice of computers and their languages requires basic concepts (strategies) of exploitation of computer capacity both from time and core memory points of view.

Differences between chemical and biochemical processes are based on the inherent complexity of the living system. To reveal this complexity with the utilisation of computers it is necessary to formulate as accurately as possible mathematical models of cellular activity and to construct algorithms for data logging and data analysis.

Mathematical models can be analysed by off-line computer application. Simulation as well as checking the results with experimental data will result in the acceptance or refinement of the model. Optimisation procedures make it possible to find the optimal environmental conditions for a particular process.

Information on the status and the dynamic behavior of a fermentation process can be obtained through on-line computer application. Data-acquisition systems serves the purpose of collection and storing of experimental data for later utilisation. On-line computer analysis of the process makes it possible to multivariate the data of physical and "gateway" sensors resulting in instantaneous availability of information on process variables of vital importance. By choosing suitable algorithms for data analysis, it is possible to increase significantly the number of observations on cellular metabolism often without an increase in the number of sensors.

Logged and reduced experimental data can be used for on-line, real-time control or can be used for off-line simulation and optimisation purposes. The most suitable tool for the purpose of on-line data acquisition and process analysis is a highly instrumented-computerised pilot-plant fermenter. The current trend in the fermentation industry is away from banks of poorly monitored fermentation vessels. Instead of these, pilot-plants consisting of one or two highly instrumented fermenters using on-line computerised data logging and analysis, are in operation.

Replacement of conventional analog controllers by Direct Digital Control systems has proven successful in cases of large-scale operation. However, it is realized that non-interacting, closed-loop-control techniques are less adequate for establishing optimal environmental conditions in batch fermentation processes. There is a requirement of mutual interactions among the control, the physical and biological state variables during process control.

An interactive control system can be specifically developed for biological processes for fermentation process control purposes. Here a master program coordinates the functions of individual controllers during the process. The control decisions are formulated in algorithms constructed on the basis of simulated, experimentally analysed data and the results of process optimisation. The biological interactive control programs can be first experimentally tried on the pilot-plant scale and then transferred to the industrial scale.

Although the application of computers in biochemical engineering is in its infancy, the early experiences indicate that the cost of installation of computers, both on pilot-plant and industrial scale is strongly compensated by the results achieved in the maximisation of yield and economy of plant operation. With the price of computers rapidly decreasing it is easy to imagine that most fermentation processes will be utilising computers in the near future.

Symbols

a_0	$(X_{initial}/X_{final})$* constant
a_1	maximum specific growth rate (μ_{max}),
A_i	a rational number the value of which is characteristic of the organism,
$b_1, b_2, ..., b_n$	parameters,
C_i	n-element vector relevant controllable variables associated with i,
C_n	vector of any control variable,
D	$(=F/V)=$ dilution rate,
F	volumetric flow rate,
$F(\gamma)$	$=(a_0 + a_1^{\gamma a_2} + a_3 e^{a+\gamma})$,
$g(P)$	objective function,
k_i	a positive number the value of which is characteristic of the environment,
$k_1, k_2, ..., k_n$	velocity constants,
K	X in stationary phase,
MU	growth rate (in the CSMP program),
$MUMAX$	maximum growth rate (in the CSMP program),
P_i	m-element vector affected by implementing alternative i,
PV	predicted value of a state variable (Fig. 18),
S	substrate (concentration),
t	time,
t_0	starting time of a process,
t_2	lag time,
T	temperature,
U	sum of $u_1, u_2, ..., u_n$, uncontrollable variables,
V	liquid volume,
v_i	value of implementing proposed alternative i,
$z_1, z_2, ..., z_n$	adjoint variables,
$x_1, x_2, ..., x_n$	vectors of state variables,
x_1	growth, (cell-mass),
x_2	product (metabolite),
X	product (cell-mass) concentration,
Y	yield.

Greek Letters

α	constant,
β	constant,
γ	$t - t_2$,
ϕ	apparent coefficient for growth activity,
μ	specific growth rate,
θ	temperature (as control variable).

References

Ackhoff, R. L.: Scientific Method: Optimizing Applied Research Decisions. New York: John Wiley, Inc. 1962.

Aiba, S., Humphrey, A. E., Millis, N. F.: Biochem. Eng., New York: Academic Press 1965.

Andrews, J. F.: Biotech. Bioeng. 10, 707 (1968).

Bacher, S., Kaufmann, A.: Ind. Eng. Chem. 62, 53 (1970).

Box, G. E. P., Hunter, W. G.: Technometrics 4, 301 (1962).

Bristol, E. H.: Proc. 25th Ann. ISA Conf. Philadelphia, Pa., No. 561 (1970).

Bungay, H. R. III: Process Biochem. 6, 38 (1971).

Burns, D. A.: Paper presented at 71st Ann. Meeting of ASM, Minneapolis, Miss. (1971).

Carnahan, B., Luther, H. A., Wilkes, J. O.: Applied Numerical Methods. New York: John Wiley and Sons, Inc. 1969.

Chen, M. S. K., Nagai, Sh., Humphrey, A. E.: Biotech. Bioeng. 13, 257 (1971).

Constantinides, A., Spencer, J. L., Gaden, E. L., Jr.: Biotech. Bioeng. 12, 803 (1970a).

Constantinides, A., Spencer, J. L., Gaden, E. L., Jr.: Biotech. Bioeng. 12, 1081 (1970b).

Cooney, C. L., Wang, D. I. C., Mateles, R. I.: Biotech. Bioeng. 11, 269 (1969).

Cotter, J. E.: Chem. Eng. Progr. 65, 52 (1969).

Deindoerfer, F. H.: Advan. Appl. Microbiol. 2, 321 (1960).

Deland, E. C.: Chemist — The Rand Chemical Equilibrium Program. Memo RM-5404-PR. Rand Corp. Santa Monica, Calif (1967).

Edwards, V. H., Wilke, C. R.: Biotech. Bioeng. 10, 205 (1968).

Edwards, V. H.: Biotech. Bioeng. 12, 679 (1970).

Edwards, V. H., Gottschalk, M. J., Noojin, A. Y., Tuthill, L. B., Tannahill, A. L.: Biotech. Bioeng. 12, 975 (1970).

Emshoff, J. R., Sisson, R. L.: Design and Use of Computer Simulation Models. New York: MacMillan Inc. 1970.

Erickson, L. E., Humphrey, A. E.: Biotech. Bioeng. 11, 467 (1969a).

Erickson, L. E., Humphrey, A. E.: Biotech. Bioeng. 11, 489 (1969b).

Erickson, L. E., Fan, L. T., Shah, P. S., Chen, M. S. K.: Biotech. Bioeng. 12, 713 (1970).

Falch, E. A., Gaden, E. L., Jr.: Biotech. Bioeng. 12, 465 (1970).

Fletcher, R., Powell, M. J. D.: Computer J. 6, 163 (1963).

Friedman, M. R., Gaden, E. L., Jr.: Biotech. Bioeng. 12, 961 (1970).

Fuld, G. J.: Advan. Appl. Microbiol. 2, 351 (1960).

Gaden, E. L., Jr.: Chem. Ind. (London) 1955, 154.

Garfinkel, D.: In: Heinmets, F. (Ed.): Concepts and Models of Biomathematics, Vol. 1, p. 1. New York: M. Dekker Inc. 1969.

Garfinkel, D., Garfinkel, L., Pring, M., Green, S. B., Chance, B.: Ann. Rev. Biochem. 39, 473 (1970).

Grayson, P.: Process Biochem. **4**, 43 (1969).
Gyllenberg, H. G., Koskenniemi, E., Rauramaa, V.: Biotech. Bioeng. **11**, 757 (1969).
Harmes, C. S. III: Paper Presented at 1971 Ann. Meeting of SIM, Fort Collins, Colorado (1971).
Harrison, D. E. F., Chance, B.: Appl. Microbiol. **19**, 446 (1970).
Hastings, J. W.: Ann. Rev. Biochem. **37**, 597 (1968).
Heinmets, F. (Ed.): Concepts and Models in Biomathematics, Vol. 1. New York: M. Dekker, Inc. 1969.
Henry, R. A.: Instrumentation Technol., p. 47 (June, 1970).
Ho, L. Y., Humphrey, A. E.: Biotech. Bioeng. **12**, 291 (1970).
Hoshi, H.: J. Soc. Instr. Control Eng. **6**, 658 (1967).
Hsieh, D. P. H., Silver, R. S., Mateles, R. I.: Biotech. Bioeng. **11**, 1 (1969).
Humphrey, A. E.: Proc. of LABEX Symposion on Computer Control of Fermentation Processes. London (1971).
IBM Corporation (Ed.): System/360 Continuous System Modeling Program. User's Manual (360A-CX-16X). H. 20-0367-1, New York (1967).
Kendall, D. G.: J. Roy. Soc. **B11**, 230, 266, 281 (1949).
Koga, S., Burg, C. R., Humphrey, A. E.: Appl. Microbiol. **15**, 683 (1967).
Koga, S., Kagami, I., Kao, I. C.: In: Perlman, D. (Ed.): Fermentation Advances, p. 369. New York—London: Academic Press 1969.
Kono, T., Asai, T.: Biotech. Bioeng. **11**, 19 (1969 a).
Kono, T., Asai, T.: Biotech. Bioeng. **11**, 293 (1969 b).
Kono, T., Asai, T.: Hakko-kagaku Zasshi **49**, 128 (1971 a).
Kono, T., Asai, T.: Hakko-kagaku Zasshi **49**, 133 (1971 b).
Lengyel, Z. L., Nyiri, L.: Biotech. Bioeng. **8**, 337 (1966).
Li, Y. T.: Proc. of First IFAC Congress, Moscow. London—Washington, D.C.: Butterworths 1960.
Luedeking, R., Piret, E. L.: J. Biochem. Microbiol. Technol. Eng. **1**, 393 (1959).
Megee, R. D., Kinoshita, S., Frederickson, A. G., Tsuchiya, H. M.: Biotech. Bioeng. **12**, 771 (1970).
Monod, J.: Ann. Rev. Microbiol. **3**, 371 (1949).
Mori, M., Yamashita, S.: Control Eng. **14**, 66 (1967).
Naito, M., Takamatsu, T., Fan, L. T., Lee, E. S.: Biotech. Bioeng. **11**, 731 (1969).
Nyiri, L.: Z. Allgem. Mikrobiol. **7**, 107 (1967).
Nyiri, L. K.: Proc. of LABEX Symposion on Computer Control of Fermentation Processes. London (1971 a).
Nyiri, L. K.: Paper Presented at 1971 Ann. Meeting of SIM, Fort Collins, Colorado (1971 b).
Nyiri, L., Lengyel, Z. L.: Biotech. Bioeng. **7**, 343 (1965).
Nyiri, L., Lengyel, Z. L., Erdélyi, A.: J. Antibiotics (Tokyo) Ser. A. **16(2)**, 80 (1963).
Nyiri, L. K., Tóth, G. M.: Biotech. Bioeng. **13**, 687 (1971).
Nyiri, L. K., Tóth, G. M., Corso, V.: Biotech. Bioeng. to be published.
Pennington, R. H.: Introductory Computer Methods and Numerical Analysis. MacMillan Co., London, 2nd Edition. p. 417 (1970).
Pirt, S. J., Callow, D. S.: J. Appl. Bacteriol. **23**, 87 (1960).
Pirt, S. J., Righelato, R. C.: Appl. Microbiol. **15**, 1284 (1967).
Pizer, S. M., Ashare, A. B., Callahan, A. B., Brownell, G. L.: In: Heinmets, F. (Ed.): Concepts and Models of Biomathematics, Vol. 1, p. 105. New York: M. Dekker Inc. 1969.
Prokop, A., Erickson, L. E., Fernandez, J., Humphrey, A. E.: Biotech. Bioeng. **11**, 945 (1969).
Ramanathan, M., Gaudy, A. F.: Biotech. Bioeng. **13**, 125 (1971).

Ramkrishna, D., Frederickson, A. G., Tsuchiya, H. M.: J. Ferment. Technol. **44**, 203 (1966).

Rhyne, V. T., Leavitt, L. A., Peterson, C. R., Canzoneri, J.: Computer Programs in Biomedicine **1(1)**, 36 (1970).

Sawaragi, Y., Sunahara, Y., Nakamizo, T.: Statistical Decision Theory and Adaptive Control Systems. New York: Academic Press 1967.

Sheppard, C. W.: FEBS Letters **28**, 14 (1969).

Shu, P.: J. Biochem. Microbiol. Technol. Eng. **3**, 95 (1961).

Starling, T. D., Pollack, S. V.: Introduction to Statistical Data Processing. Prentice-Hall Inc., Englewood Cliffs, N.J. (1968).

Storey, C., Rosenbrock, H. H.: Computation Methods in Optimization Problems. New York: Academic Press 1964.

Tanner, R. D.: Biotech. Bioeng. **12**, 831 (1970).

Theiss, D. J., Hobbs, L. C.: Datamation, May 15, p. 25 (1971).

Tsai, B. I., Erickson, L. E., Fan, L. T.: Biotech. Bioeng. **11**, 181 (1969).

Tuffile, C. M., Pinho, F.: Biotech. Bioeng. **12**, 849 (1970).

White, J., Hazbun, E. A.: Proc. 25th ISA Conf., Phila., No. 511 (1970).

Wimpenny, J. W. T.: In: Microbial Growth. Proc. of 19th Symposium of Soc. for Gen. Microbiol. London-Cambridge: Cambridge University Press 1969.

Wyatt, P. J.: Nature **226** (5242), 277 (1970).

Yamashita, S., Hoshi, H., Inagaki, T.: In: Perlman, D. (Ed.): Fermentation Advances, p. 441. New York-London: Academic Press 1969.

Yano, T., Koga, S.: Biotech. Bioeng. **11**, 139 (1969).

Young, T. B.: ACS National Meeting, Div. Microbiol. Chem. Technol. No. 39, Chicago (1970).

Young, T. B., Bruley, D. F., Bungay, H. R. III: Biotech. Bioeng. **12**, 747 (1970).

Zajic, J.: J. Ind. Microbiol. **5**, 326 (1964).

Zines, D. O., Rogers, P. L.: Biotech. Bioeng. **12**, 561 (1970).

L. K. Nyiri, Ph. D.
Technical Director
Fermentation Design, Inc.
Bethlehem, PA 18017/USA

CHAPTER 3

Mixed Microbial Populations

ANTHONY F. GAUDY, JR. and ELIZABETH T. GAUDY

With 11 Figures

Contents

1. Introduction

The terms "mixed microbial populations", "natural populations", and "heterogeneous populations" have all been used to denote systems in which natural selection occurs and determines which organisms can survive and which will predominate in the ecosystem. From time to time, shifts in predominating species may occur, due to natural interactions between species or in response to an external stimulus. The external stimulus may be some change in the chemical, physical, or biological environment which affects the system and to which the population responds, but which is not brought about internally by the population itself. Thus a natural microbial ecosystem is indeed an enrichment culture which tends to seek whichever balance or imbalance of species can evolve under the existing environmental or cultural conditions. Such systems occur everywhere in nature, and they are also employed under somewhat controlled (or unnatural) conditions in various biological engineering processes. The most important use of such populations is in the biological treatment of wastes, and it is this application which is the primary concern of this chapter.

Biological treatment processes for organic wastes are designed to alleviate undue stress on the natural bodies of water which serve essentially as reservoirs and, to some degree, as sites of purification for used water. Undue stress upon receiving waters may be defined as any stress which would tend to create anaerobic conditions in these receiving bodies, or to upset the "natural" ecology of a healthy stream. Thus, the primary purpose of biological treatment processes is to remove overloads of organic matter which would disturb the natural ecological balance by affecting the desired aerobic environment in the water course. Such treatment processes employ, almost exclusively, mixed microbial populations cultured under aerobic conditions.

It is important to emphasize some of the similarities and differences between the biological treatment of wastes and other bioengineering operations. In addition to the use of heterogeneous populations, rather than selected cultures, in waste treatment, little or no selection of substrates is possible. The substrates are generally whatever organic compounds are present in the waste water to be treated. Little or no tailoring of this proposed growth medium is practiced, other than addition of nitrogen and phosphorus when the waste is deficient in these nutrients. Toxic substances must be removed, diluted, or otherwise neutralized prior to treatment. In some cases the pH of the waste may be adjusted, but precise pH control is seldom employed. The reactors are usually open to the elements, and temperature is not controlled. Sterilization of the medium and the air supply is obviously not practiced. In the most general case the harvested product consists of the overflow from a gravity settler (clarifier). The efficiency of separation depends on various factors, the most important of which is the degree of agglomeration or flocculation of the heterogeneous population developed in the growth reactor, which may in turn depend to a great extent upon the types of organisms included in the population. The clarity of the supernatant is also dependent in large measure upon an ecological relationship in which free-swimming predators (largely protozoa) are required to complete the clarification process by ingesting dispersed or nonflocculated bacterial cells.

Thus, numerous differences can be recognized between the biological waste treatment processes and the usual industrial (commercial) fermentations. Considering all of these differences and the potential for disruption of steady controlled reaction which they represent, one might question whether any real quantitation or predictive evaluation can be applicable to such processes. Although the heterogeneity of the population represents a source of intrinsic internal variation in the system which adds greatly to its comparative complexity and the difficulty of understanding and controlling it, the situation is not as chaotic as it may at first seem. There are basic similarities to more controlled fermentations, since the principles of aeration and agitation, overall growth kinetics of aerobic organotrophic organisms, etc., apply, regardless of the species involved. The process is in some respects more simple than many industrial fermentations, because the aim is simply to remove the organic substrates. It is in some respects more complicated, since both metabolic and ecological responses can occur when the cultural conditions change, and these systems are subjected to such changes in greater degree and with much greater frequency than are commercial fermentations.

The degree of operational control installed and used in these plants (as in any fermentation) is usually governed by the value of the product. The value of usable water is only now being realized; therefore more sophisticated control devices and techniques should be economically justifiable in the future, provided their use is scientifically justified, i.e., that their use really leads to increased efficiency and reliability. It is therefore important to determine which operational controls are really needed and which are not. Obviously, sterilization of the "medium" and the air supply is an unneeded precaution, and may even be detrimental to the process. Close pH and temperature control may not be necessary in view of the fact that some amount of variation in these parameters, while fostering changes in the ecology of the system, may not interfere with the objective of the process — removal of the organic carbon sources. Changes in ecology can, after all, come about regardless of maintenance of close control over such parameters as pH and temperature.

It is not incumbent upon the bioengineer to attempt precise control of the biological process simply for the sake of control, but it is necessary to control the process to a degree consistent with reliable delivery of a plant effluent of acceptable quality for return to the water resource (the receiving stream) or for subsequent treatment by other unit processes prior to direct reuse. Such control will require a thorough understanding of the kinetics of biological purification of waste waters and of the mechanisms of substrate removal in mixed microbial population systems.

In discussing some aspects of recent progress toward understanding biological processes of waste treatment, attention will be focused primarily upon processes employing aerated, completely mixed, fluidized reactors, i.e., completely mixed activated sludge processes. The advantages of reactors of this type for processes which must be closely controlled are familiar to those who design and operate industrial fermentation processes. The probability that standards of acceptability of waste treatment effluent quality will become more rigid, that increased demand for water will increase its value, and that closer control of waste treatment processes will become economically feasible supports the expectation that reactors of this type will become increasingly important in treatment of wastes. While it is not presently possible to achieve consistently the desirable degree of control with any biological waste treatment process, it appears that completely mixed fluidized reactors offer the most fruitful approach toward controlled production of a high quality effluent. It might be argued with some validity that the ideal biological treatment process would be one which required no operational control. Unfortunately, the idea that such processes can be designed is a delusion, and efforts devoted to achieving this "technological Utopia" could be more profitably directed toward determining the types of controls needed, and devising ways and means to effectuate them.

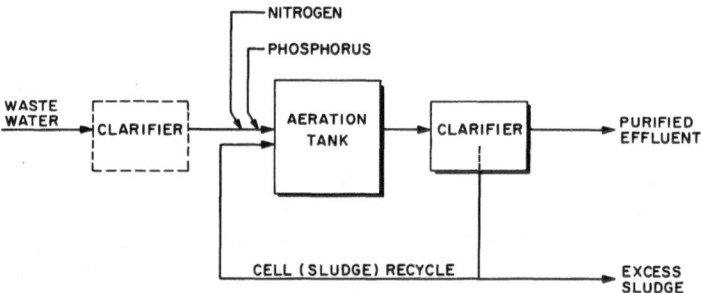

Fig. 1. Flow diagram for typical activated sludge process

A simplified flow diagram for an activated sludge process is shown in Fig. 1. The aerobic reactor (aeration tank), clarifier, and sludge recycle line are intrinsic features of activated sludge processes. Municipal wastes contain suspended matter (inorganic and organic) in quantities sufficient to warrant their separation by gravity (primary treatment) prior to admitting the waste water to the aeration tank-clarifier (secondary treatment). For many industrial wastes such pre-treatment is not necessary, but other pre-biological treatment steps may be desirable. Many industrial wastes require nutrient supplementation, i.e., addition of nitrogen (NH_4^+) and phosphorus (PO_4^{3-}), and sometimes adjustment of pH.

The purified effluent from the clarifier may be channeled to the receiving stream or may be subjected to one or more of a variety of further polishing ("tertiary") treatments dictated by the needs and immediate uses of the effluent. Several processes are available for disposal of excess sludge, all of which have attendant problems.

In this chapter we shall examine some recent developments pertinent to the behavior of mixed microbial populations in the aerobic reactor of completely mixed activated sludge processes. Also some aspects of response of such systems to changes in the environment will be presented. Some applications of this knowledge in the development of new and innovative processes for waste treatment and for alleviation of the sludge disposal problem will be scrutinized.

2. Substrate Removal and Growth

The primary aim of aerobic biological treatment of wastes is removal of organic carbon. The assimilation of organic substrates leads to an increase in the mass of the biological population, and if a source of nitrogen (as well as other required nutrients) is present, cell replication can occur. Whether one considers growth as replication or simply as an increase in mass, the phenomenon occurs as a consequence of accomplishment of the primary aim of the biological treatment process, and the excess "sludge" becomes a waste product which then presents a disposal problem. Partially because of this problem and also because a good deal of the biological engineer's knowledge is gleaned from the field of microbiology, as much attention is usually focused on defining growth kinetics as on the primary concern, substrate removal. It would be ideal if one could devise a system so delicately balanced that new growth would be compensated for by an equal amount of death and decay (to CO_2 and H_2O) so that there was no net increase in the mass of the population. In such a case the organic matter in the waste could be totally oxidized to the relatively non-pollutional endproducts CO_2 and H_2O. This intriguing possibility will be discussed in a later section. However, most processes for biological treatment of wastes do involve an increase in the mass of the population, i.e., net microbial growth.

a) Methods of Measurement

It is desirable, before describing studies of mixed microbial populations, to discuss methods of measurement of the parameters employed. Since these methods are applied to systems in which both the organic substrates and the microbial species are heterogeneous, variable and essentially undefinable, methodology must necessarily differ in some respects from that employed in defined, pure culture systems. The two parameters of primary concern are substrate removal (waste purification) and microbial growth (sludge accumulation or increase in biological solids).

The determination of total organic carbon as COD (chemical oxygen demand) is the method of greatest utility for measurement of substrate removal. (For a description of the method of measuring COD, see *Standard Methods for the Examination of Water and Waste Water*, 1965). In studies using actual wastes, the constituents of which are not known, only a measurement of total organic substrate could be employed. Even in laboratory experiments using a defined medium, excretion of partially oxidized products of metabolism often occurs (Krishnan and Gaudy, 1965). Measurements of substrate removal based on specific analyses for the original substrate result in false estimates of purification in such cases. It is, therefore, preferable to measure COD in all cases, even though more specific analyses may also be made.

Measurements of microbial growth must also utilize methods appropriate to the population under study. Gravimetric measurement has the widest applicability and is, of course, the method of greatest accuracy. Viable counts can be used only in a few types of laboratory experiments. Activated sludge in a treatment plant or in the usual laboratory pilot plant is flocculated to an extent which precludes the use of viable counts. Optical density measurements may be used in cases where flocculation is minimal, if restricted to the range where optical density is proportional to mass, and can be converted to mass using a calibration curve.

In general, it is preferable in all work dealing with waste purification by the activated sludge process to utilize the methods of measurement which are applicable to all such studies in the laboratory or in field installations, i.e., COD and biological solids measured gravimetrically. (In treatment plants where appreciable quantities of inorganic suspended matter may be present, biological solids are measured by gravimetric determination of the total suspended solids less ash content and expressed as volatile suspended solids).

b) Hydraulic Control of Microbial Growth

Most texts on microbial physiology present the familiar growth cycle curves beginning with the lag phase and continuing through the logarithmic phase, the declining growth phase, the stationary phase, and the death phase for a batch system initiated with an ample amount of carbon source and other nutrients and a loopful of cells. Most attention has been focused on the logarithmic growth phase which finds mathematical description in the differential form:

$$\frac{dN}{dt} = \mu N \tag{1}$$

where N represents the number of cells per unit volume, or

$$\frac{dX}{dt} = \mu X \tag{2}$$

where X represents the mass of cells per unit volume. The latter equation, for reasons discussed above, is more generally applicable and will be employed herein.

The specific growth rate, μ, is best considered as an inverse function (equal to 0.693/doubling time) of the time required for a doubling of the population regardless of the parameter used to measure growth, i.e. numbers, mass, total protein, etc. In the logarithmic growth phase this number is constant. The doubling time for a particular system may be, for example, one hour or three hours, but regardless of the value, if the doubling time is constant, the system is exhibiting logarithmic growth. For any specific organism under a specific set of cultural conditions (type of carbon source, temperature, constituents of the medium), there exists a maximum specific growth rate, μ_{max}, when no nutrient is present at a growth-limiting concentration. This value is characteristic of the organism and μ_{max} is a valuable parameter for assessing the effect of different cultural conditions on a single organism or for comparing different organisms under the same cultural conditions. It is important, however, to avoid the frequent error of considering μ_{max} to be "the" logarithmic growth rate constant.

The engineering utility of μ as a gross kinetic parameter becomes more apparent when we consider continuous culture systems rather than batch systems. In such systems the hydraulic or mixing regime plays as important a role in the culturing of mixed microbial populations as it does in more defined aerobic "fermentation" systems. Most reactors can be made to approach the condition of complete mixing more easily than one of plug flow. In reactors approaching complete mixing with respect to all contents (soluble and insoluble), the concentration of microbial cells exiting the reactor is the same as that in the reactor. Thus, as mixed liquor is displaced from the reaction vessel by incoming medium containing few or no cells, the total mass of cells in the reactor is decreased or diluted at the rate of $X(F/V)$; i.e., X is decreased by a factor equal to the ratio of the flow rate, F, to the aeration liquor volume, V. This ratio is defined as the dilution rate, D. The rate of change in the concentration of cells for such a system (once-through, i.e., no cell recycle) is given below:

$$\frac{dX}{dt} = \mu X - D X .\qquad(3)$$

Defining a "steady state" condition as one in which $dX/dt = 0$, it is apparent that

$$\mu = D .\qquad(4)$$

Thus, in a continous culture at steady state, the specific growth rate is equal to the dilution rate, while the biological solids level, X, is determined by the concentration of the limiting nutrient. For systems employing cell recycle, μ is not equal to D but can be related to it in a definable fashion. Thus, in continuous culture systems, with or without cell recycle, the biological parameter μ, i.e., the logarithmic growth rate constant, is subject to hydraulic control. This relationship is as theoretically valid (no more and no less) for heterogeneous or mixed populations as it is for pure culture growth systems.

There is, however, some uncertainty regarding the possibility of maintaining a steady state with respect to X, i.e., $dX/dt = 0$, when mixed microbial populations are employed. There may indeed be room for argument concerning the precision with which a true steady state can be maintained for pure cultures. However, in systems in which there is opportunity for interplay not only between various types of bacteria but between bacteria and their predators as well, gross changes in predominating species can occur in either cyclical or random patterns, and there is ample cause for questioning the validity or the practical utility of the concept of a steady state with respect to cell concentration. Since various species exhibit different efficiencies of conversion of substrate to cellular material, and since the efficiency of this conversion may not be the controlling factor in determining the dominance of species at any given time, the concentration of cells (the sludge concentration) in the reactor may exhibit considerable variation. If there were considerable variations in X, and if these caused like variations in the amount of substrate in the reactor (organic carbon in the effluent), the mathematics of the steady state might not be useable in design and operation. Data are available (Gaudy, Ramanathan, and Rao, 1967) which indicate that there is indeed enough fluctuation in X to warrant a conclusion that a steady state can only be roughly approximated, but the fluctuation in substrate concentration in the reactor is less than that in cell concentration. Since it is substrate concentration which is of paramount concern in waste treatment, steady state mathematics can be used.

Formerly, activated sludge processes were analyzed, in the main, as plug flow processes, and kinetic and mechanistic analyses of the process were based largely upon the ideas and concepts of the typical batch system growth cycle. Thus, it may be important to emphasize that when one considers completely mixed fluidized culture processes (e.g., a completely mixed activated sludge system) and employs the kinetics of the steady state, he must realize that the system is maintained in a logarithmic growth phase. In order for the cell concentration to approach a steady state ($dX/dt = 0$), the doubling time (i.e., $0.693/\mu$ or $0.693/D$) must remain constant, defining the condition of logarithmic growth.

c) Sludge Production, or Cell Yield

While μ and, indirectly, N and X, are related to the hydraulic regime, these biological parameters are affected by chemical and biochemical factors as well. The most important for our purposes are those factors related to assimilation of the carbon source, i.e., to the purpose of biological treatment. In any bacterial system, one may state that the rate of accumulation of biological solids, dX/dt, is proportional to the rate of utilization of the organic substrate, dS/dt. The proportionality factor can be defined as the cell yield, Y, thus

$$\frac{dX}{dt} = -Y\frac{dS}{dt},$$

or

$$Y = -\frac{dX}{dS}.$$

(5)

The above equation implies that regardless of the kinetics of the system, ΔX is related to a larger but proportionally constant increment, ΔS. Thus, if one measured Y in the log phase or in the declining phase of growth, the same value would be obtained. This statement is usually taken to be true, but should be subjected to more experimental verification than is usually the case. Fig. 2 shows a plot of substrate removed (expressed as chemical oxygen demand, COD) vs. biological solids concentration for a mixed microbial population growing on glycerol. Data points to the left of the dotted vertical line indicate samples taken prior to the end of the logarithmic growth phase. It can be seen that Y remains essentially constant regardless of the phase of growth in which it is measured. Cell yield is considered to be a biological "constant" and for an individual growth experiment this seems justified. Insofar as its measurement for a particular experiment is concerned, the usual procedure of measuring total growth and amount of carbon source removed at the point of maximum growth seems valid. However, care must be taken not to generalize from one experiment to another when dealing with yield values for heterogeneous populations. The result of one yield determination on a specific carbon source is not often of real predictive value for

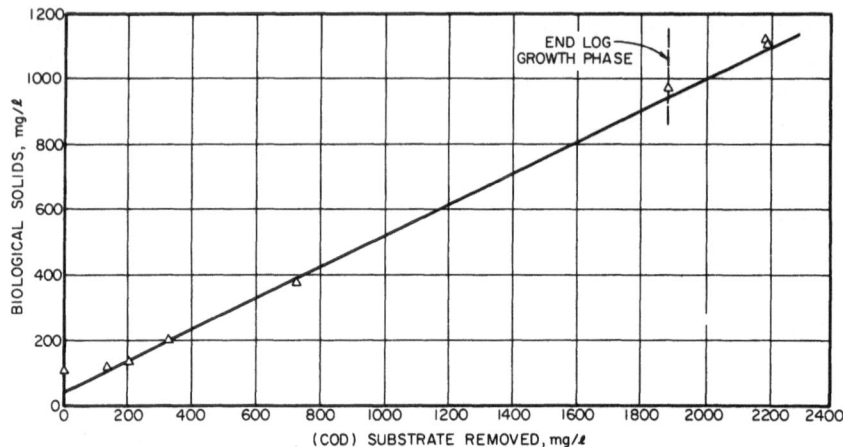

Fig. 2. Relation between biological solids production and substrate consumption throughout growth of a mixed microbial population on glycerol. Y, the slope of the line, is essentially constant

the next experiment on the same carbon source. Although cell yield does appear to be relatively independent of the growth rate or phase of growth, it is not independent of species since the efficiency of conversion of carbon source to cellular components is not the same for all species. Some workers have attempted to provide rather precise estimates of Y based upon thermodynamic properties of the carbon source (Servizi and Bogan, 1963). Such oversimplifications can lead to misinterpretation by design engineers, since they attempt to override the unique variable in mixed microbial systems, i.e., the changing predominance of the species.

It is expedient to treat Y as one of the system constants in attempts to depict the operational behavior of an activated sludge process. It is, however, necessary to temper one's use of the term "constant". For example, an analysis of 118 separate determinations of sludge yield on glucose for heterogeneous populations of sewage origin growing in batch systems resulted in an average value of 61.9%. The range of values, however, was from 36 to 88%. The coefficient of variation for the mean from this rather large number of samples was 20%. The 95% confidence limit for the mean was ± 2.3 (Gaudy and Ramanathan, 1971).

d) Growth Rate and Concentration of Limiting Nutrient

The specific growth rate, μ, is known to be related to the concentration of the growth-limiting nutrient in batch cultures. As in the case of Y, we will again consider the organic substrate as the limiting nutrient, although any required nutrient, for example, nitrogen source, can exert a controlling effect on μ. In the case of biological treatment of some industrial wastes which contain very small amounts of nitrogen or phosphorus, these constituents are usually added in amounts large enough to ensure that they do not limit either μ or total growth on the given amount of organic carbon source.

Various relationships between μ and S have been examined. It is often assumed in designing waste treatment plants that there is a linear relationship ($\mu = kS$) up to a concentration at which μ becomes independent of S, i.e., up to μ_{max}, under a given set of growth conditions. Other relationships have been proposed by Monod (1949), Moser (1958), and Teissier (see Schultze, 1964). An examination of these relationships for fit with experimental data for mixed microbial populations indicated that the hyperbolic function described by the Monod equation provided the most adequate expression of the dependence of μ on S for these populations (Gaudy, Ramanathan, and Rao, 1967). The Monod equation for μ as a function of S, is given below:

$$\mu = \frac{\mu_{max}(S)}{K_S + (S)} \tag{6}$$

where K_S is the substrate concentration at which $\mu = \mu_{max}/2$.

The values of K_S and μ_{max} are constant for a pure culture under a given set of cultural conditions, but vary with different organisms under identical conditions.

The value of the saturation constant, K_S, determines the curvature of the plot of μ vs. S; high values yield very flat curves, and low values very sharp curves (see Fig. 10, Gaudy, Ramanathan, and Rao, 1967).

The similarity between Eq. (6) and the Michaelis-Menten equation for relating the velocity of an enzymic reaction to substrate concentration is immediately apparent. The various ways of plotting the data which are employed for the Michaelis-Menten equation can also be applied to the Monod expression. One of the straight line forms of the equation is given below:

$$\frac{S}{\mu} = \frac{S}{\mu_{max}} + \frac{K_S}{\mu_{max}}. \tag{7}$$

A plot of S/μ vs. S permits evaluation of μ_{max} as the reciprocal of the slope of the straight line plot and K_S as the product of μ_{max} and the value of the intercept on the S/μ axis. However, if one's data plot as a straight line in accord with Eq. (7), this does not provide evidence of a hyperbolic relationship between μ and S. The curvilinear nature of the relationship can best be observed from a plot of μ vs. S. A recommended procedure in analyzing such growth data is to make a plot of μ vs. S and sketch in a curve through the data. Then, using one of the slope intercept forms [e.g., Eq. (7)] determine values for μ_{max} and K_S. Substituting these values into Eq. (6), the values of μ for various substrate concentrations can be calculated and the resulting curve can be compared with the observed data.

The similarity in form of the Monod and the Michaelis-Menten equations may make it tempting to draw a close analogy between the "Michaelis-Menten constant", K_M, and the saturation constant, K_S, as well as between maximum enzyme velocity and maximum growth rate. This should be avoided because, while the Michaelis-Menten equation for enzyme kinetics can be defended on the basis of mechanistic theory, the Monod relationship is entirely empirical and simply provides a convenient way of relating observed values of μ obtained at various values of S. The saturation constant is probably best envisioned as representing no more than can be gleaned from the analytical geometry of the hyperbolic curve generated by Eq. (6). It is the numerical value of S when μ is equal to $0.5 \mu_{max}$. Since the curve becomes flatter as K_S increases, the saturation constant might be considered as some quantitative expression of the sensitivity of μ (or of the population, expressed through the observed μ) to the concentration of limiting nutrient, S, as μ approaches the maximum value, μ_{max}, for the system under study.

It is also important to note another difference; in the Monod expression it is the logarithmic growth rate constant (specific growth rate), not (as in the Michaelis-Menten equation) the velocity or rate of reaction, which is related to S. It should also be pointed out that in enzyme studies the velocity of the reaction is determined under conditions which assure that zero order (not first order) kinetics prevail. In a study of enzyme kinetics, the amount of active substance (enzyme preparation) causing the kinetic is held constant, whereas in a growth study the catalytic agent (the microbial population) acting upon the substrate is increasing — logarithmically. Add to this the fact that in a growth study the substrate must pass into the cells, that it may be subject to a concentration effect inside the cell and that many (not one) interdependent reactions are occurring, and the unrelatedness of the two equations is even more apparent. Superimpose on these facts the ecological considerations necessary in a growth experiment employing mixed microbial populations, i.e., the fact that the observed increase in overall population can involve interrelationships between species, and the fallacy of assuming a close tie (other than the analytical geometry of the expressions) between the Michaelis-Menten and Monod equations should be apparent.

If one were to set up a series of growth flasks, each containing the same volume of medium but a different amount of substrate and an equal inoculum of microorganisms, the series of growth curves generated would (for the straight line portion on a semilogarithmic plot) exhibit increasing slopes for flasks containing increasing concentrations of substrate. Increasing initial concentrations of substrate would produce decreasing increments in μ until at some concentration μ would reach a maximum value, μ_{max}. The value of μ plotted vs. the substrate concentration S would yield a plot which can be described by Eq.(6). There are ample data in the literature indicating that this expected result does indeed occur (Schaefer, 1948; Monod, 1949; Gaudy, Ramanathan, and Rao, 1967).

While μ can be shown to be dependent upon S in batch experiments of the type described above, it is necessary to consider the effect of changes in S upon μ before deciding upon the use of the Monod equation or any model relating specific growth rate, μ, and substrate concentration, S. These considerations should serve to emphasize again the empirical and approximate (rather than exact) nature of relationships between μ and S.

Let us consider Eq.(6) which states that when S changes, μ changes. Let us also consider that this equation was first devised as a means of relating observed μ values to various substrate concentrations initially present in batch growth experiments. The determination of μ in a batch experiment involves plotting the logarithm of N or X vs. time throughout the growth period. The specific growth rate is the slope of the straight line portion of this semilogarithmic plot. It should be apparent that substrate must be consumed in order to register enough "growth" so that the straight line portion can be identified and its slope determined. Thus, from a strictly theoretical standpoint, each change in S cannot lead to a change in μ, since μ by definition is a constant describing a phase of growth during a period of time when S is changing. Thus, the symbol for substrate in the Monod equation might best be considered as initial substrate concentration, S_0. If one assumes that the Monod equation is an exact expression of the dependence of μ upon S, then growth at values of S below those at which μ is independent of S (below μ_{max}) would have to be measured during a period when S did not change. If μ changed instantaneously in response to a change in S, there would be no period of constant doubling, i.e., no logarithmic growth phase, and μ could be measured only as a tangent to a curve. The fact remains, however, that the logarithmic growth phase does exist and represents simply an experimentally measured period during which the average doubling time is constant. It makes no difference whether one measures the slope of the straight line portion of the growth curve at the beginning ("initial growth"), middle, or end of this period. One is dealing with experimental data, and the accuracy with which the straight line portion of the plot can be identified is determined by the accuracy and number of plotting points one has obtained.

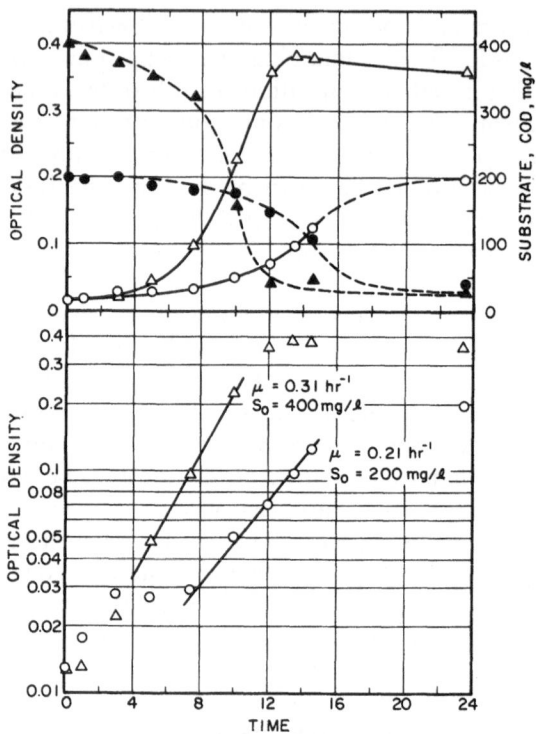

Fig. 3. Effect of initial substrate concentration on logarithmic growth rate. At both substrate concentrations, approximately 50% of the substrate was removed during the logarithmic growth phase

An examination of data obtained in batch experiments with mixed microbial populations illustrates the lack of instantaneous response of μ to changes in S. Fig. 3 shows results for two of the concentrations of substrate (glucose at 200 and 400 mg/l) used in a growth experiment. All reactors were seeded with small equal portions of the same heterogeneous microbial population developed on glucose from an initial seed of municipal sewage. It would be preferable to have more data points than were obtained, but the arithmetic plots of substrate concentration and optical density appear to be adequate for kinetic assessment of the course of growth and substrate removal. It is seen from the semilog plot of optical density that a reasonable estimate of μ can be made, and that μ was dependent upon initial substrate concentration; i.e., the higher initial substrate concentration allowed a higher specific growth rate. Both values of S shown were below that at which μ_{max} was developed, so that S was theoretically limiting at all times during growth. The point to be emphasized here is that in both reactors a reasonably well-defined exponential phase of growth was developed. In order for this phase to be developed and measured experimentally, substrate had to be consumed. An examination of the data shows that during the logarithmic phase, i.e., during the period when μ was constant, at these concentrations, which were initially below the substrate concentration for development of μ_{max}, approximately 50% of the organic substrate was removed.

Turning attention again to Eq.(6), it should become apparent that the mathematical equality expressed is only approximately consistent with the observed behavior of biological systems which the equation was originally devised to depict (i.e., it was originally a curve-fitting exercise in which the hyperbolic curve of Eq.(6) could be passed through experimentally determined values of μ for different initial concentrations of substrate). We must accept, then, the idea of "slippage" in the Monod relationship. Such a realization does not destroy its usefulness; indeed, it may in the long run enhance it because it emphasizes the empirical nature of the relationship. In view of the complexity of the growth process it would be an entirely over-simplistic view to expect the logarithmic growth constant (specific growth rate) to obey, precisely, the equality stated in Eq.(6). To paraphrase a well-known quotation (Orwell, 1946), all equalities are equal but some are more equal than others. From these experiments and others like them, as well as the discussion above, we may conclude that μ is somewhat dependent upon S and is related to S in a manner which is in general accord with the hyperbolic function of Monod [Eq.(6)]. The dependence is not as tightly coupled as Eq.(6) would seem to indicate; specific growth rate, μ, is not so easily disturbed or changed by a change in S as the equation would lead one to expect. And this fact is one which it is well to remember when analyzing the effect of shock loadings of organic matter in biological waste treatment reactors (see below).

Values of μ_{max} and K_S are often determined in batch experiments as described previously, but they may also be determined in continuous culture experiments. In continuous flow, one may run a once-through completely mixed reactor at increasing dilution rates, and the value of D at which the microbial population approaches zero and the substrate concentration in the reactor tends to that in the feed can be taken as the value of μ_{max} if K_S is small in relation to S. If one substitutes D for μ in Eq.(6) and solves for K_S, the following equation is obtained:

$$K_S = \frac{S(\mu_{max} - D)}{D}. \tag{8}$$

Operating the reactor at a value of D less than that at which the culture is washed out, the observed steady state value of S could be inserted into the equation and a value of K_S determined. Determination of μ_{max} and K_S in continuous culture is a more time-consuming operation than determination of these biological "constants" in batch studies. There are advantages and disadvantages for measurement of the "constants" in either type of system. Sometimes the value of K_S may be so low as to make its measurement from batch data extremely difficult. On the other hand, evaluation of μ_{max} in continuous flow by the method described above is somewhat difficult because of the S-shaped or "tailed-off" dilute-out curve which is sometimes observed even in pure culture studies and may be expected to occur with natural populations for which changes in dilution rate represent an obvious selective pressure fostering changes in predominating species and, as a consequence, changes in μ_{max}. Values of μ_{max} and K_S from batch data for mixed microbial populations are also subject to variation due to ecological changes.

Values of μ_{max} and K_S (at 25° C) have been determined for heterogeneous microbial populations of sewage origin selected by growth on specific carbon sources. A considerable amount of data has been obtained using glucose as source of carbon. It should be emphasized that the heterogeneity of the population makes it as unreasonable to expect constant values for μ_{max} and K_S as it is for cell yield, Y. Thus, approximate values or a useful range of values must be accepted if one is to relate μ and S. Values of μ_{max} and K_S for growth on glucose have generally fallen in the ranges 0.5—0.6 hr^{-1} and 50—125 mg/l, respectively (Gaudy, Ramanathan, and Rao, 1967; Ramanathan and Gaudy, 1969).

The maximum logarithmic growth rate constant, μ_{max}, is generally attained at substrate concentrations in excess of 400—600 mg/l. Such con-

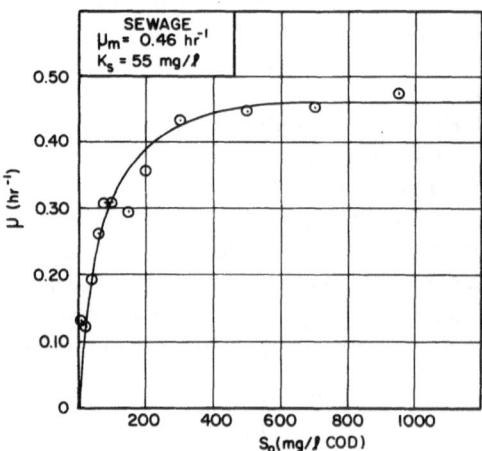

Fig. 4. Monod plot for mixed microbial population of sewage origin growing on the soluble fraction of municipal sewage (Peil and Gaudy, 1971)

centrations of organic matter are available in some industrial wastes, but municipal wastes seldom attain such high organic concentrations, at least for the portion normally subject to biological treatment, i.e., the effluent from the primary clarifier. Thus, μ_{max} cannot be assessed directly in a single batch experiment, and it is difficult to obtain a range of S values for growth experiments. In a recent study (Peil and Gaudy, 1971), municipal waste water was concentrated under vacuum and the soluble portion was employed in batch growth studies to examine the effect of S on μ. Data obtained in one such study are shown in Fig. 4, wherein values of μ obtained at various concentrations of S are plotted. The data describe a hyperbolic curve characteristic of the Monod relationship. The values of μ_{max} and K_S, which were obtained from a plot of S/μ vs. S, were within the range of those generally observed with glucose as substrate.

e) Utility of the Monod Equation for Mixed Microbial Populations

At this point it may be concluded (tentatively, at any rate) that the growth "constants" Y, μ_{max}, and K_S, which have so long been employed in pure culture growth studies, can be employed for heterogeneous populations, but that the constants are not so very constant and we must settle for a range of values or for approximate values. We have violated the sanctity of Eq. (6) by the demonstration of the loose linkage or slippage in the response of μ as S changes, but this violation might be expected to hold for pure culture studies as well as for those with mixed microbial populations. Despite this "weakness" in the equation and the impossibility of obtaining constant values for the kinetic constants for mixed microbial populations, the Monod relationship is a very valuable one for kinetic description of batch and continuous flow systems and can be usefully employed in equations of the "steady state" for description of the behavior of continuous flow reactors.

3. Kinetic Equations for Continuous Culture with Recycle

It was shown algebraically above that, in the steady state in continuous culture reactors, $\mu = D$. The hydraulic regime in completely mixed reactors is such that no other conclusion is tenable. According to Eq. (6), μ is also related to S. The early papers on continuous culture by Monod (1950) and by Novick and Szilard (1950) dealt with these important relationships and laid the foundation for kinetic description of continuous culture. These works and the expert elaboration and extension of them by Herbert and his coworkers (Herbert, Elsworth, and Telling, 1956; Herbert, 1961) have provided general kinetic "theory" and served as the basis for much useful research, in both basic and applied areas.

Since all mixed microbial (activated sludge) systems for purifying waste involve the recycling of biological solids, the steady state equations given by Herbert for systems employing cell recycle are particularly pertinent to these systems. These equations are given below:

$$\bar{S} = \frac{K_S(A)(D)}{\mu_{max} - AD},$$ (9)

$$\bar{X} = \frac{Y}{A}(S_i - \bar{S}),$$ (10)

$$A = (1 + a' - a'b).$$ (11)

In these equations, S_i is the concentration of organic substrate in the feed solution, a' is the volumetric feedback ratio, i.e., the ratio of feedback flow to feed medium flow. The concentration factor, b, is the ratio of the cell concentration (X_R) in the feedback flow to the cell concentration in the reactor, $\bar{X}$. Both a' and b are operational constants selected by the investigator. These hydraulic parameters and the inflowing substrate concentration, S_i, together with the biological constants μ_{max}, K_S, and Y, control $\bar{X}$ and $\bar{S}$ at any selected dilution rate, D.

Rather long-term studies in which heterogeneous populations of sewage origin were grown in continuous culture on glucose have indicated that Eqs. (9), (10), and (11) can provide useful kinetic description for a completely mixed heterogeneous system (Ramanathan and Gaudy, 1969). However, since the heterogeneity of the population allows changes in predominance (and consequently in sludge yield, Y), the biological solids concentration in the reactor was subject to significant fluctuations (under relatively constant feeding conditions), and operation holding b constant was difficult to achieve. Also, it can be reasoned that the change in X_R required to hold b constant when X changes, fosters further change in "$\bar{X}$" rather than providing a dampening effect on variations in X $\left(\text{since } b = \frac{X_R}{X}\right)$. From a practical standpoint, it would seem, therefore, that operation holding X_R, rather than b, constant could decrease fluctuation in "$\bar{X}$" and provide a means of adapting the kinetic theories of Monod, Novick and Szilard, and Herbert and coworkers for use in design and operation of activated sludge systems.

In present-day operation there are some attempts to hold the aeration solids concentration (X) at or near a pre-selected value by varying the recycle flow (i.e., a'). Such a procedure could conceivably cause serious stress to the system. For example, consider a case wherein for some reason X began to decrease. The decrease in X might forewarn of an increase in S, i.e., a decrease in the operational efficiency of the plant. It would be desirable to increase X on the essentially correct assumption that a higher biological solids concentration provides a proportionately greater number of microorganisms for assimilation of the substrate. However, both the cell concentration and the mean residence time in the reactor $\left(\bar{t} = \dfrac{1}{(1 + a')\,D} \right)$ contribute to the efficiency of substrate removal. Now, if in order to increase X, reactor detention time were decreased (as would be the case if a' were increased) one of the factors contributing to efficiency of substrate removal would be pitted against the other. If X had to be increased, it would be better to do so by exerting control over X_R without changing reactor retention time. In any case, attempting to remedy an operational upset by direct manipulation of X is somewhat akin to "locking the barn door after the horse is stolen". It would be far better to provide some operational means of decreasing fluctuation in X.

It seems apparent that in any kinetic model, the amount of cells in the recycle must be defined and that this factor must be incorporated into the operational (and design) equations. Experience with the use of a constant concentration factor b in laboratory pilot plants has indicated its undesirability in this regard, and it may be safely assumed that even greater difficulty would be experienced in attempting to employ this mode of operation in field installations. Using X_R rather than b as an operational parameter seems, therefore, more practical, and such a procedure would contribute a greater stability with regard to the "steady state" in X for heterogeneous microbial systems. It has been previously pointed out that X will tend to fluctuate because of changes in species predominance, in any event. Also it should be re-emphasized that the purpose of biological treatment is not (at least as it is presently constituted) to culture microbial populations, but to remove relatively low concentrations of organic substrates from used waters. Although, from the viewpoint of kinetic theory it is desirable to demonstrate and maintain a steady state with respect to X $\left(\text{i.e.,}\ \dfrac{dX}{dt} = 0 \right)$, it is not necessary to do so in order to assure efficient operation with respect to substrate removal. It would seem far better to allow X to fluctuate within reasonable limits, and to attempt to gain operational control over the process by maintaining constant (or nearly so) the amount of "substrate removing material" (biological solids concentration, X_R) fed to the reactor.

The concentration of biological solids in the recycle flow is not necessarily a measure of its potential for substrate removal, since the unit activity of sludge may vary. Some of the organic solids are "dead" cells (incapable of assimilating substrate), and actually, the amount of dead material represents an added substrate loading to the system. It may have a possible advantageous effect as a weighting factor which assists in the flocculating and settling process after the mixed liquor leaves the aeration tank. Also, the "dead" material can enhance heterogeneity and diversity in microbial populations which might otherwise tend to be restricted to a narrower range of species diversity due to the nature of the carbon or nitrogen source.

In any event, a higher recycle sludge concentration can be expected to provide a larger number of substrate "removers" or "feeders". For a given sample of activated sludge there is, in general, a direct relationship between the time required to remove a given amount of soluble substrate and the sludge concentration brought in contact with the waste under aerobic conditions. While such relationships appear to remain linear over a fairly wide range of sludge concentrations, the proportionality factor varies even for growth on a specific carbon source or organic waste because of differences in the species of microorganisms comprising different sludge samples (Rao and Gaudy, 1966).

In addition to maintaining X_R constant, it is advisable to maintain a rather high concentration of recycle solids. A recycle solids concentration of approximately 10000 mg/l is not difficult to attain by gravity thickening. Values of a' of approximately 0.25 are commonly used in field installations. Thus, biological solids concentration in the reactor due to recycled cells alone might be expected to amount to about 2000 mg/l. Therefore, for a waste with S_i of 1000 mg/l, assuming a sludge (cell) yield of 50%, the expected value of $\bar{X}$ would be approximately 2500 mg/l. Such values for $\bar{X}$ in activated sludge aeration tanks are not uncommon. The fact that values of S_i are usually lower than 1000 mg/l serves to emphasize the point herein delineated, i.e., most of the biological solids in the reactor at any given time are those which have been "fed" to it in the recycle. The ratio of "feeders" to available substrate is rather high; therefore, regardless of variation in X due to growth in the reactor, the fluctuation in reactor solids concentration is minimized by maintaining X_R high, and $\bar{X}$ is steadied by maintaining X_R constant. Use of a constant X_R thus allows a closer operational approach to the ideal steady state described by kinetic theory.

Having provided arguments for using X_R rather than b as a constant for heterogeneous microbial population systems, it is appropriate to examine the consequences of such a decision on the expressions for $\bar{X}$ and $\bar{S}$. Eqs. (12) and (13) given below were developed by writing materials balances for X and S in the steady state including the terms employed by Herbert (1961), but omitting b and treating recycle cell concentration, X_R, as a system constant (Ramanathan and Gaudy, 1971). It was assumed that substrate in the recycle is of negligible concentration in the useful range of dilution rates.

Solving for $\bar{X}$ in terms of $\bar{S}$ leads to the following equation:

$$\bar{X} = \frac{Y[S_i - (1+a')\bar{S}] + a'X_R}{(1+a')}. \tag{12}$$

Solving for $\bar{S}$ leads to the quadratic form:

$$\bar{S} = \frac{-b \pm \sqrt{b^2 - 4ac}}{2a} \tag{13}$$

where

$$a = [\mu_{max} - (1+a')D], \tag{14}$$

$$b = D[S_i - (1+a')K_S] - \frac{\mu_{max}}{(1+a')}\left[S_i + \frac{a'X_R}{Y}\right], \tag{15}$$

$$c = K_S(D)S_i. \tag{16}$$

It is interesting to compare the values obtained for $\bar{X}$ and $\bar{S}$ as D varies when employing Eqs. (9) and (10) (b=constant) with those obtained using Eqs. (12) and (13) (X_R=constant). Such a comparison has recently been made by Ramanathan and Gaudy (1971) using the following values for the biological "constants": $\mu_{max} = 0.5 \, \text{hr}^{-1}$, $K_S = 75 \, \text{mg/l}$, $Y = 0.6$. A value of 0.25 for the hydraulic recycle ratio, a', was selected, since this is one commonly employed in field installations; a recycle concentration, X_R, of 10000 mg/l and a concentration factor, b, of 4 were selected, since these are values which might be readily attained in practice. The value of S_i selected was 1000 mg/l. The computed dilute-out curves in Fig. 5 show that up to a dilution rate of $1.0 \, \text{hr}^{-1}$ both models predict comparable levels of $\bar{X}$ and $\bar{S}$. As D is increased beyond this value, the system employing X_R as an operational constant manifests greater ability to resist cell dilute-out and substrate leakage.

It is particularly noted that either system can (theoretically) provide excellent substrate removal at dilution rates which might normally be considered as applicable to activated sludge processes, i.e., D values lower than $0.5 \, \text{hr}^{-1}$. Although very good substrate removal is attainable at higher dilution rates, it is somewhat inadvisable to attempt to employ higher D values for an activated sludge process of the usual type, (i.e., one employing the general flow scheme shown in Fig. 1). It is wise not to decrease the reactor detention time so severely as to leave too little time

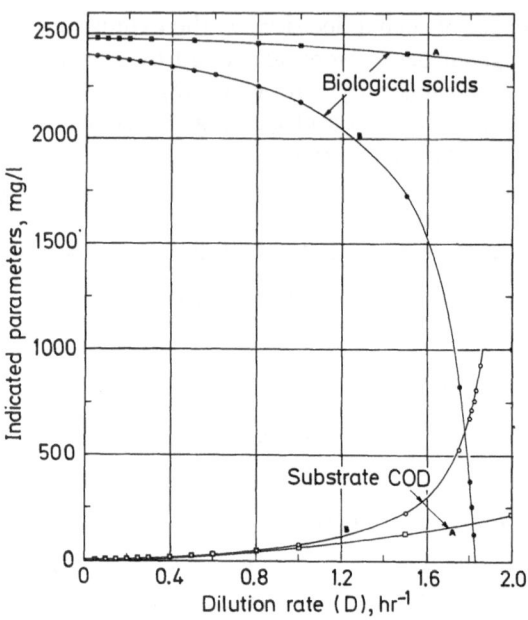

Fig. 5. Comparison of dilute-out curves for biological solids concentration and substrate concentration in completely mixed reactors operated at a constant recycle solids concentration, $X_R = 10000$ mg/l (curves A), and at a constant cell recycle ratio, $b = 4.0$ (curves B). In both cases, $S_i = 1000$ mg/l, $\mu_{max} = 0.5$ hr^{-1}, $K_S = 75$ mg/l, $Y = 0.6$, $a' = 0.25$ (Ramanathan and Gaudy, 1971)

for favorable metabolic reaction to various changes in the environment (i.e., shock loading), nor does it seem advisable to operate at D values in excess of μ_{max}. Operation at values of D below μ_{max} would seem advisable in view of the high degree of uncertainty and opportunity for periodic increases in substrate concentration inherent in activated sludge processes. However, it might be surmised from Fig. 5 that the system could conceivably handle much higher organic loadings, i.e., higher values of S_i.

Using only the model employing X_R as an operational parameter and varying both S_i and D, the behavior of the system with respect to $\overline{X}$ and $\overline{S}$ as S_i is increased can be computed and the results shown graphically. Fig. 6 shows that, at dilution rates of 0.5 hr^{-1} and below, the system provides extremely good efficiency of substrate removal at rather high values of S_i. Also, throughout the range of S_i values, the biological solids concentration in the reactor lies within a range (2000—4000 mg/l) not uncommon in field installations.

This kinetic model predicts steady operational behavior of activated sludge processes (under rather idealized conditions — perfect settling,

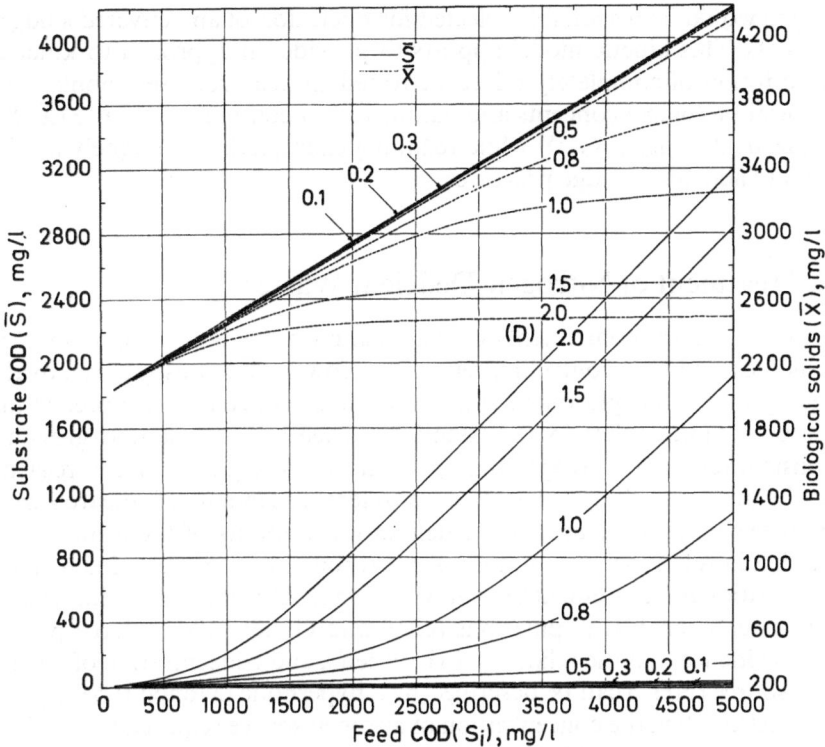

Fig. 6. Effect of inflowing substrate concentration, S_i, at various dilution rates on $\bar{X}$ and $\bar{S}$. $X_R = 10000$ mg/l, $a' = 0.25$, $\mu_{max} = 0.5$ hr^{-1}, $K_S = 75$ mg/l, and $Y = 0.6$ (Ramanathan and Gaudy, 1971)

steady environmental conditions, and constancy of the "constants"). It should be mentioned here that some of the effects of changing values of the biological constants μ_{max}, K_S, Y, and the operational parameters a' and X_R have been examined computationally, and at X_R values such as the one herein employed, the changes in the other parameters do not appear to negate the utility of the model or the steadiness imparted to the system by maintaining a high concentration of biological solids in the recycle flow (Ramanathan and Gaudy, 1971). It would seem (rather fortunately) that with activated sludge much of the unsteadiness in biochemical efficiency which could be imparted to the system because of variations in the biological parameters can be considerably dampened out by the recycling of large numbers of cells to the reactor.

It is apparent that the employment of a constant X_R and assumption of negligible substrate concentration in the recycle flow are unrealistic at very high dilution rates. However, such values of D are above the range which would be reasonably selected for operation of an activated sludge process. This kinetic model appears to provide an approach to kinetic description of completely mixed activated sludge processes, employing general biological concepts and parameters of continuous culture ($\bar{X}$, $\bar{S}$, S_i, D, a', X_R, μ_{max}, K_S, Y) while retaining characteristics uniquely pertinent to activated sludge processes.

4. Treatment of Nitrogen-Deficient Wastes

Of the chemical constituents of the medium other than carbon source which can exert a controlling effect on growth and substrate removal, nitrogen and phosphorus are the most critical, since they are needed in larger quantities than other essential nutrients. For balanced growth, considerably more nitrogen than phosphorus is required, and in regard to the activated sludge process, a considerable amount of research has been expended to determine guidelines for amounts of these nutrients needed in relationship to amounts of carbon source present (for review of nutritional requirements, see Sawyer, 1956). Recent studies with heterogeneous microbial populations (Goel and Gaudy, 1969a) have provided evidence that the relationship between μ and concentration of nitrogen source (NH_4^+) can be described by a hyperbolic (Monod type) equation. Also, when the concentration of nitrogen source required for attaining μ_{max} was compared to the concentration of carbon source required to reach μ_{max}, the ratio of carbon source to nitrogen source was found to be in approximate agreement with the guideline values given by Sawyer (1956). However, it is not necessarily desirable to base the selection of a particular ratio of carbon source to nitrogen source on requirements for optimum growth. Again, the primary aim of biological treatment is to remove the carbon source(s) and the same degree of purification efficiency can be attained at different COD:N ratios (and different sludge nitrogen contents), depending upon other variables, such as dilution rate (Goel and Gaudy, 1969b).

A concern which has assumed increasing importance is the amount of inorganic nitrogen in effluents from treatment plants. In a case wherein nitrogen must be added to the waste in order to treat it, a subsequent leakage of nitrogen is an economic as well as a pollutional concern. One can lower the amount of supplemental nitrogen needed for a particular degree of treatment (removal of carbon source) by increasing the reactor detention time (i.e., decreasing the dilution rate). On the other hand, decreasing the dilution rate for a given ratio of carbon to nitrogen source can foster leakage of nitrogen in the effluent. Such relationships have recently been shown for heterogeneous microbial populations of sewage origin grown in completely mixed reactors (Goel and Gaudy, 1969 b). In general, it would seem that the usual practice of providing a ratio of biochemical oxygen demand (BOD) to available nitrogen source of 20:1 for nitrogen-deficient industrial wastes is somewhat wasteful of nitrogen.

It has been known for many years that soluble organic substrates can be assimilated in the absence of a nitrogen source by various microorganisms (Siegel and Clifton, 1950). Using heterogeneous populations, it was shown some years ago (Gaudy and Engelbrecht, 1963) in batch experiments at an initial ratio of carbon source to biological solids concentration of approximately 1.0, that the rate and extent of both sludge accumulation and removal of carbon source were nearly the same in the absence of a source of nitrogen as in the presence of an ample amount of available nitrogen. The characteristically low concentrations of substrate in the incoming waste in comparison to the biological solids concentration is a feature of activated sludge systems which requires special attention, since it affects not only the kinetics of the purification process, but the mechanism of substrate removal as well. Rao and Gaudy (1966) presented experimental data and a hypothesis which stated, in brief, that if the biological solids were made sufficiently high, the rate of substrate removal should approach linearity, because the mechanism of removal was primarily oxidative assimilation of the carbon source into non-nitrogenous cellular materials (largely stores) even in the presence of an ample amount of nitrogen source for balanced growth conditions. This work was extended by Krishnan and Gaudy (1968), who were able to show that with high biological solids concentration the mechanism of substrate removal was essentially the same whether surplus nitrogen source or no nitrogen source was present, or when an inhibitor of protein synthesis was present. Thus, assimilation of the carbon source was not dependent upon the presence of a source of nitrogen.

The effect of biological solids concentration on the kinetic and mechanistic behavior of heterogeneous populations under both growing and nonproliferating conditions is readily discerned in Figs. 7 and 8 (Thabaraj and Gaudy, 1971). In these batch experiments, both the substrate removal and subsequent endogenous phases are shown. During the substrate removal phase shown in Fig. 7 (COD:biological solids = 20), protein synthesis predominates, whereas in the substrate removal phase in Fig. 8 (COD:biological solids = 3), the situation is reversed. A considerable difference in both the rate and mode of kinetic expression under growing and nonproliferating conditions is evident in Fig. 7, while the kinetic and mechanistic behavior in the two systems with and without a nitrogen source is much more alike at the higher initial concentration of biological solids (Fig. 8).

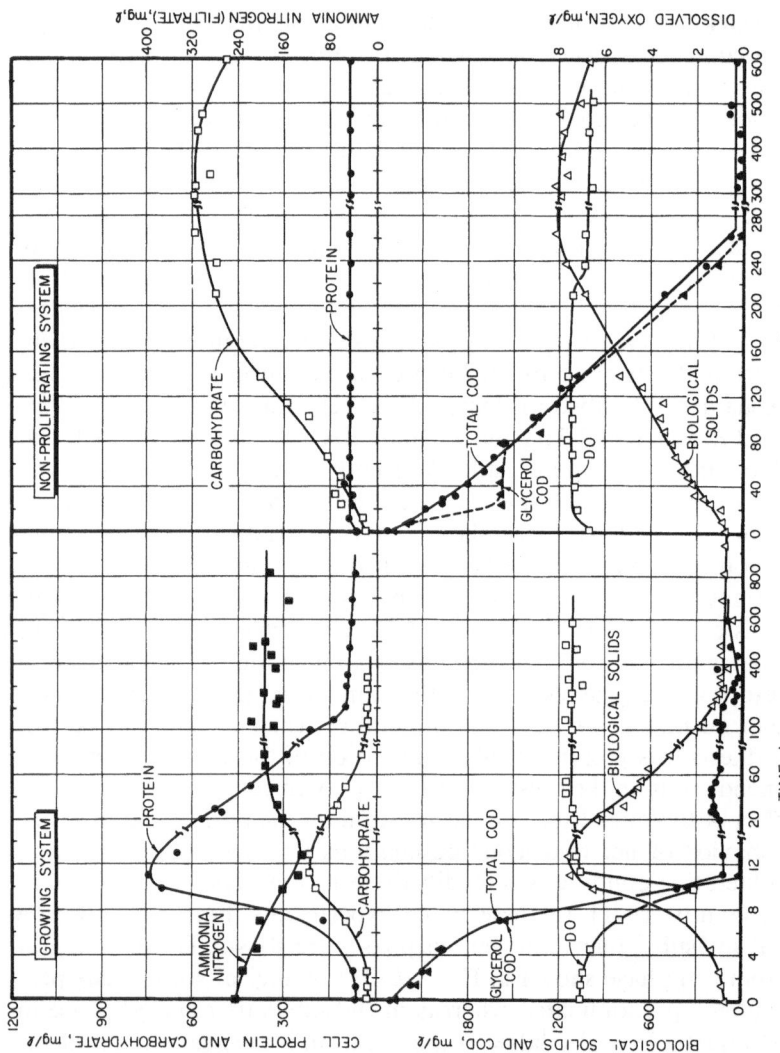

Fig. 7. Comparison of substrate removal and endogenous phases for mixed microbial populations under growing and nonproli-ferating conditions for initial COD : biological solids ratio of 20 : 1 (Thabaraj and Gaudy, 1971)

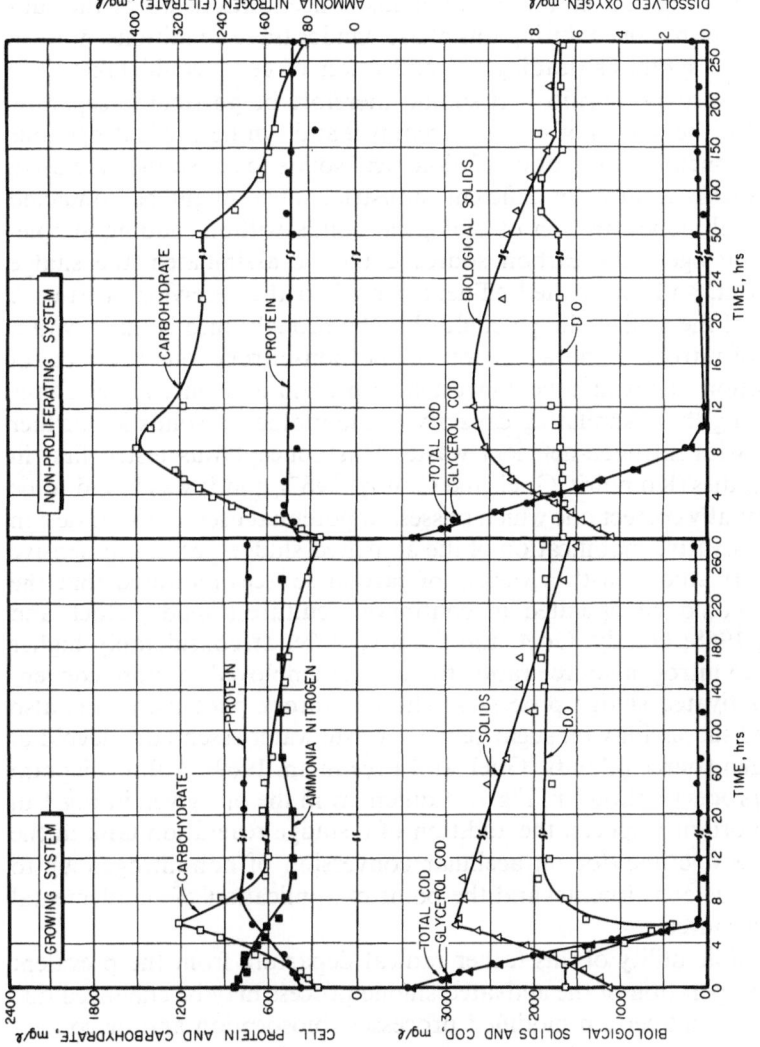

Fig. 8. Comparison of substrate removal and endogenous phases for mixed microbial populations under growing and nonproliferating conditions for initial COD : biological solids ratio of 3 : 1 (Thabaraj and Gaudy, 1971)

One might surmise that the metabolic assimilation and synthesis of non-nitrogenous stores of carbon and energy for subsequent channelling into protein synthesis and subsequent replication is a process characteristic of cells placed in fresh medium, e.g., cells recycled to the aeration tank, regardless of the relative concentrations of carbon source and cells, and that at very high initial biological solids concentrations all of the substrate can be removed during this phase. Under these conditions, most of the protein synthesis (leading to replication of cells) could take place after the waste is "purified". It should therefore be possible to separate, physically, the phases of substrate removal and synthesis of nitrogenous cell components if high ratios of biological solids to substrate were used. For example, a nitrogen-deficient industrial waste might be contacted with a high concentration of biological solids without adding supplemental nitrogen; the carbon source could be assimilated, the sludge settled out, and the treated effluent passed to the receiving stream. A portion of the settled sludge could then be brought into contact with a source of nitrogen, and a portion of the non-nitrogenous products of assimilation might then be converted to protein and nucleic acid, thus revitalizing the assimilating capacity of the biological solids for further contact with nitrogen-deficient waste. This concept was tested first in batch studies (Komolrit, Goel, and Gaudy, 1967), and it was found to be an essentially correct one which possessed potential usefulness for design of an innovative modification of the activated sludge process. Extensive laboratory studies with a variety of carbon sources indicated that the process could be operated in continuous culture (Gaudy, Goel, and Gaudy, 1968; Gaudy, Goel, and Gaudy, 1969) at considerably higher carbon to nitrogen source ratios than those employed in more conventional activated sludge processes. The laboratory pilot plant was also operated successfully on sugar refinery waste water essentially devoid of a nitrogen source (Gaudy, Goel, and Freedman, 1969). A flow diagram for the process is shown in Fig. 9. It differs from the one given in Fig. 1 in two important respects: the addition of a sludge reaeration tank in the sludge recycle line (for intracellular conversion of non-nitrogenous to nitrogenous constituents) and the point of application of supplemental nitrogen source.

The possible utility of this rather radical departure from the prevalent mode of operation of the activated sludge process may be enhanced due to the fact that certain modified processes (biosorption and/or contact stabilization plants) make use of a sludge reaeration tank. In these plants the nitrogen source is added to the incoming waste in the conventional manner, as in Fig. 1. The theory upon which this type of treatment is based assumes that the mechanism of waste purification involves "sorption" of the carbon source during a short contact time with a very high

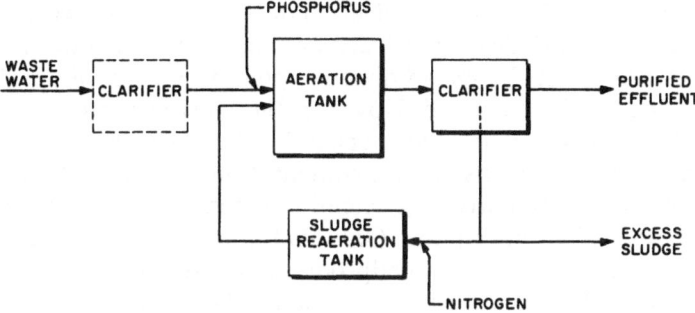

Fig. 9. Flow diagram for proposed oxidative assimilation process for treatment of nitrogen-deficient industrial wastes

concentration of activated sludge and subsequent "burning-off" (oxidizing) of these materials in the sludge reaeration tank. Thus, while the flow sheet for a biosorption plant is similar to that shown in Fig. 9, except for the point of addition of nitrogen, the proposed mechanism of treatment is quite different. The metabolic behavior of sludge contacted with a carbon source at a high solids:COD ratio, which has been amply documented as described above, predicts that much of the nitrogen added to the waste before contact with the sludge would be wasted in the biosorption process. Since little synthesis of nitrogenous material would occur during the short contact time, a large percentage of the added nitrogen would be carried into the receiving stream with the effluent where it would contribute to algal growth. Such plants may, with some possible adjustment in retention time in the contact and reaeration tanks, be converted to the mode of operation shown in Fig. 9. This mode of operation is intended especially for nitrogen-deficient industrial wastes, and it appears to offer an innovative alternative which can effect significant savings in the cost of supplemental nitrogen, while safeguarding against leakage of the added nitrogen.

5. Disposal of Sludge

a) Separate Disposal

An integral part of the biological waste treatment process is a means of disposal of excess biological solids. As stated previously, one of the reasons for the importance of defining the kinetics of growth, as well as substrate removal, is that cell replication does occur during waste purification; i.e., excess sludge is produced and must be treated further. In a few cases, sludge may be spread directly upon the land for disposal, or it may be dried and incinerated. However, the most common method of disposal involves either aerobic or anaerobic digestion which reduces the volume of material and its organic content. Both aerobic and anaerobic digestion can involve fairly long detention times, and neither represents an ultimate method of disposal. Only that portion of the sludge which can be converted to gaseous endproducts is actually removed, and a large volume of particulate matter remains. This is usually dried and used as a soil conditioner, or it may be incinerated or disposed of by other methods.

Both digestive processes are interesting from the viewpoint of microbial ecology, since rather complex microbial interactions may be assumed to occur. The anaerobic process has been used for many years, both for disposal of untreated, settled sewage solids and for disposal of excess microbial cells from secondary biological treatment. The mixed microbial population responsible for anaerobic digestion includes anaerobic bacteria capable of hydrolyzing large molecules and fermenting them to low molecular weight products, e.g., acids or alcohols, which serve as substrates for the very specialized methane-forming bacteria. If the process is to function properly, a very delicate ecological balance between the varied fermentative organisms and the methane bacteria must be maintained. Frequent digester failures attest to the ease with which this ecological balance can be upset. The process requires continual sensing and analysis of liquid and gaseous phases if operational control is to be attained. However, in spite of much excellent research, there is still insufficient understanding of the process to indicate what parameters should be used for monitoring and what control measures are most effective.

In addition to the undigested sludge which remains and requires disposal, the liquid effluent from anaerobic digestion contains rather high concentrations of organic matter (endproducts of fermentation) which require recycling of the effluent to the aerator or other further treatment. In view of these difficulties, continued widespread use of the anaerobic digestion process for sludge seems unlikely. However, anaerobic digestion of certain concentrated wastes for conversion of large molecules to smaller fermentation products, followed by aerobic treatment of these products, or possibly use of the products, would seem to offer interesting possibilities in the future technology of waste recycle processes.

A more recently popular method for disposal of sludge is aerobic digestion, in which settled sludge is simply aerated for a period of up to 10—20 days. Endogenous utilization of cellular material and predatory activity within the mixed microbial population reduce the volume of sludge before ultimate disposal. While this process avoids many of the difficulties inherent in the anaerobic process, it does not offer the ultimate solution to the sludge disposal problem. As mentioned previously, it would be most desirable to convert all of the organic matter in the waste to CO_2 and water, thus in effect operating without net synthesis of cells. The process described below offers an approach to this ideal situation.

b) Total Sludge Recycle

Some years ago in studies on the biological treatment of skimmed milk wastes, Porges, Jasewicz, and Hoover (1953) concluded that the decrease in biological solids concentration due to endogenous respiration could, if the aeration period were sufficiently extended, balance the increase due to synthesis of new cells. Therefore they proposed that a system could be made to operate, with no wasting of sludge, at an equilibrium concentration of biological solids. The incoming waste could thus be treated with the same degree of efficiency of substrate removal as with a conventional process, but the organic matter removed would all be oxidized (respired) to CO_2 and H_2O, i.e., totally oxidized. The flow sheet for such a plant would be similar to that shown in Fig. 1, except that all settled solids would be returned to the aerator. Their conclusion and proposed process (total oxidation or extended aeration activated sludge) were not generally accepted. Although many extended aeration plants are in operation, the general conclusion of a number of researchers (e.g., Symons and McKinney, 1958; Kountz and Fourney, 1959; Busch and Myrick, 1960) has been that the biological tenets upon which the process was proposed are theoretically unsound. The general view has been that the process cannot function biochemically without some wastage of biological solids (either purposely or inadvertently). Arguments against the use of the process may involve the idea that a microbial cell cannot "endogenate" itself to CO_2 and water, or that many microbes produce extracellular material (largely polysaccharide) which they do not later oxidize.

While these statements are correct, they are not completely relevant to mixed microbial populations. Although for convenience the biomass or sludge is sometimes considered as an entity or the cells are considered as "average" or "composite" organisms, the internal ecology of an activated sludge is complex. It is not unreasonable to expect that waste products of one cell can be food for others, or that cells can be lysed biologically and the walls, membranes, and internal constituents metabolized by other cells. For example, examination of the curves for biological solids and total COD in the growing system of Fig. 7 reveals that essentially all of the biomass synthesized in the substrate removal phase was oxidized in the subsequent endogenous phase (note the return of biological solids concentration to the initial value with no rise in soluble organic matter, i.e., chemical oxygen demand, COD).

Long-term operation of a laboratory scale activated sludge with total cell recycle (controlled positively by centrifugation of the mixed liquor) under continuous flow conditions has shown that biological solids do not follow a continually rising trend. There were prolonged periods of accumulation followed by periods wherein the autodigestive process exceeded synthesis, and total solids concentration decreased. (Autodigestion is used here to indicate collective activity of the biomass.) Such cyclic phenomena depend upon a changing succession of predominating species. While this system seldom operated at a steady equilibrium solids concentration as suggested by Porges *et al.*, neither did it build up a high fraction of inert solids nor did it exhibit gradual loss of purification efficiency (Gaudy, Ramanathan, Yang, and DeGeare, 1970; Gaudy, Yang, and Obayashi, 1971).

Such studies demonstrate that a process for the total oxidation of waste without net sludge synthesis is theoretically possible. However, the length of periods of solids accumulation cannot be predicted or controlled, and it might be expected that at times the sludge concentration could build to such an extent as to hinder its separation in the settling tank. At such times it is possible to provide a "chemical assist". Experimental evidence has been obtained (Gaudy, Yang, and Obayashi, 1971) which suggests the utility of the process shown in Fig. 10. It seems rea-

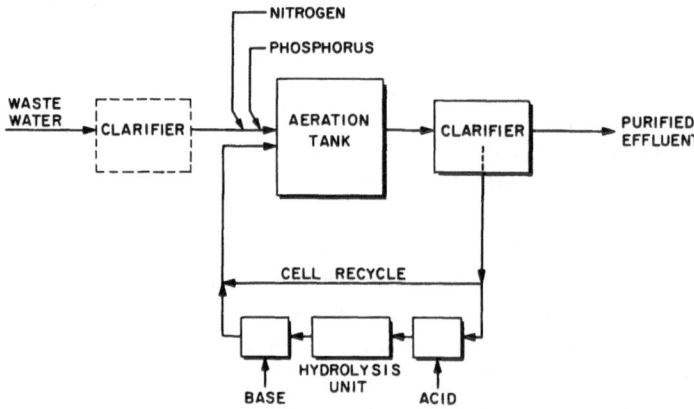

Fig. 10. Flow diagram for proposed extended aeration process with "hydrolytic assist"

sonable to expect that, of the autodigestive steps, the most difficult to initiate metabolically are those preparatory to assimilation, i.e., the hydrolysis of insoluble cell components. Therefore, if one were to assist the system by hydrolyzing a portion of the sludge (when necessary) and recycling the solubilized organic matter as shown in Fig. 10, one could gain positive engineering control of the sludge concentration. The process outlined in the figure makes use of both chemical and biological treatment to enhance controlled operation of a total oxidation process. It does chemically a function which is biologically difficult to perform, and does biologically a function which would be costly to accomplish by purely chemical means.

6. Response to Changes in Environment

Biological engineers active in the commercial fermentation industries are well aware of the effect of cultural (physical, chemical, biological) conditions on the quality and quantity of biological products and the economics of producing them. Much effort is usually expended to determine "optimum" cultural (environmental) conditions and to accommodate them in the design and subsequent operation of the process. Considerably less effort in this regard is expended on biological waste water treatment processes. They are designed on the basis of some "average" condition, and the expectation (or hope) is that they will function within a reasonably acceptable range of efficiency with regard to removal of the carbon sources in the waste.

In actual operation, there will be considerable fluctuation around the "average" conditions. Activated sludge processes, because of the use of heterogeneous microbial populations, the nature of the substrate(s), variations in volumes of material, etc., are more akin to "natural" biological processes and therefore more accountable to the vagaries of nature. Changes in the environment involving such factors as temperature, pH, hydraulic rate of feed, concentration of carbon source, type of carbon source, etc., are the rule, not the exception, in nature and, in large measure, in biological treatment plants. Therefore, insights gained in studying responses of activated sludge processes to such changes should help toward understanding occurrences in natural bodies of water (the receiving streams) which are indeed "in-stream" biological reactors which work in conjunction with the "on-shore" reactors and processes of the biological engineer. One cannot overemphasize the depth and scope of this vital area of investigational concern or the scientific and technological harvest which continued penetration of it can yield.

Both drastic and subtle changes of the types listed above have been the subject of rather extensive study in the authors' laboratories. Due to the extent and complexity of these investigations as well as space limitation, they will not be summarized here; however, it can be stated that all of the environmental changes listed above can lead to disruption of the orderly growth process envisioned in any "steady state" kinetic model. An attempt will be made herein to delineate some of the basic parameters which appear critical to depicting transients initiated in response to changes in substrate concentration and composition.

Even if the external environment is held constant, the heterogeneity of the population and the opportunities for species interaction create "internal" environmental changes. So it is quite improper to adopt the concept of precise conformity with steady state behavior for mixed microbial populations. Even with respect to biological solids concentration, such a state $\left(\dfrac{dX}{dt} = 0\right)$ is not attained, but it does not actually have to be attained for proper operation. Indeed, it is the small transients in the growth response due to minute changes in substrate concentration which formed the basis for the steady state continuous culture theories of Monod (1950) and of Novick and Szilard (1950). However, when environmental changes are imposed which cause significant deviation from the pseudo steady state with respect to substrate $\left(\dfrac{dS}{dt} = 0\right)$, leading to a significant increase in S during the transient state between initial and final steady states, the system has undergone a functional failure and the malfunction is said to have occurred because the process could not accommodate the "shock loading" which was imposed on it.

a) Quantitative Shock Loads

Let us consider briefly an environmental change consisting of an increase in concentration of carbon source, S_i, with no change in D. If $\bar{S}$ is to remain steady $\left(\dfrac{dS}{dt} = 0\right)$, then at the first small increase in S, μ would increase slightly, increasing X, which in turn would cause S to return to the steady state value; therefore, μ would return to its steady state value, i.e., D. Through repetition of this sequence with small incremental increases in S, X would undergo an observable transient increase to its new steady state level for the new S_i, and S would remain at its previous steady state level. Such a desirable response would be adjudged to be a successful accommodation of the shock loading. Changes of significant but not large magnitude can be accommodated. For example, at $D = 1/8 \ \text{hr}^{-1}$, a 50% increase in S_i might be expected to be accommodated without a significant transient in S. However, with the system operating at the same dilution rate, a larger change in S_i can be expected to cause an increase in S. In this case, substrate concentration in the reactor has been increased more rapidly than the specific growth rate, μ, can increase to accommodate it; the increase in X cannot keep pace with the rising S. Eventually, the biological solids concentration will "catch up", the rising trend in substrate concentration will be diminished, and S can be expected to return (approximately) to the initial steady state level. Implicit in the above analysis of the behavior of X and S during the transient response is the fact that the we have made use of some sort of "relationship" between μ and S. Under steady operation, the relationship that has been shown to provide a reasonably good fit to a large quantity of experimental data is Eq. (6), previously introduced. It is natural, then, to question its descriptive or predictive usefulness when rather large changes in S_i occur.

Some years ago, Perret (1960) presented theoretical kinetic arguments which led him to postulate that in an environment in which S was increasing, the observed μ would lag that predicted by the Monod equation, and in the reverse case it would lead. He termed this postulated phenomenon "growth rate hysteresis". Storer and Gaudy (1969), reasoning solely from considerations of metabolic control mechanisms, felt that Perret's theory was essentially correct, and were able to present experimental evidence for the occurrence of growth rate hysteresis in mixed microbial populations. Young, Bruley, and Bungay (1970) have used these data and data from other studies of transient responses in various pure cultures to develop a mathematical model which may more closely approach description of the dynamic behavior of continuous flow systems during transients. As previously noted, the experiments shown in Fig. 3 also show that μ and S are not necessarily as tightly coupled as would appear from the mathematical equality implied by the Monod equation or, perhaps, by any other equation relating μ and S. These are by and large empirical formulations, and one should not be surprised when the Monod equation does not provide a precise trace of the transient response in X and S when S_i is sharply increased. The model was not designed to consider the effects of sudden, large changes of any type. However, varying degrees of imprecision may be acceptable or, in any event, may have to be accepted for models describing biological treatment with mixed microbial populations. This is because complexities introduced for reasons of population heterogeneity and also the heterogeneity in the nature of the substrate can exert effects not predictable by any gross kinetic model. It seems essential, therefore, that we make use of these valuable tools (the models) with full cognizance of their approximate nature and inadequacies for certain situations.

From the analysis given above, it may be discerned that S increased because X did not increase rapidly enough to hold S to its steady state level prior to the shock. In an engineering sense, one could therefore provide an assistance by increasing X_R. Such a procedure could be facilitated if one were operating the system using the steady state model (Ramanathan and Gaudy, 1971) in which X_R rather than b is used as a "system constant". To increase X_R in response to a predetermined amount of increase in S or S_i would require constant, automated monitoring of S or S_i or both (as COD), automatic control of X_R, and possibly a reserve store of biological solids. Such operational procedures might be possible in the future, although much more sophisticated knowledge regarding biological responses is needed before such sophistication in operational apparatus can be justified. On the other hand, one can purposefully maintain a high concentration of X_R, i.e., one considerably in excess of that actually needed to provide the required or desired S at the design dilution rate. Armed with such reserve substrate assimilating capacity, the biomass can be expected to handle or attenuate a rise in S due to an increase in S_i. Thus, there is another argument, in addition to that previously presented, for providing means for positive control of X_R. Although this is seldom done, it can be said that the biological solids concentration in most activated sludge systems is much in excess of that needed to remove the substrate in the retention time provided, and this helps to add stability to the process. The usual practice of recycling rather high concentrations of biological solids is not necessarily based upon the reasoning outlined above. These solids, after leaving the aeration tank, are separated from the mixed liquor by gravity settling. The flocculating and settling characteristics of the "sludge" are somewhat affected by the solids concentration, and the seperability of sludge and effluent liquor is related to the maturity of the ecological community. Too low (or, for that matter, too high) a concentration of solids could militate against separation of the solids and this, of course, could be just as severe a functional failure as a metabolic or biochemical malfunction. In general, it can be said that the biological solids concentrations usually needed for proper separation in the clarifier are in excess of those needed for substrate removal. Thus, there are various reasons for maintaining X_R rather high in relation to the incoming feed.

Thus far in analyzing response to changes in S_i, X has been considered as a biological entity; i.e., μ has been treated essentially as an *en masse* response of the microbial population. This approach is necessary for certain purposes, but it is somewhat oversimplistic and is not wholly satisfying even for a "pure" culture which can exhibit considerable diversity of the population in response to selective pressures. For a heterogeneous population, a sudden change in environment (a shock) exerts a

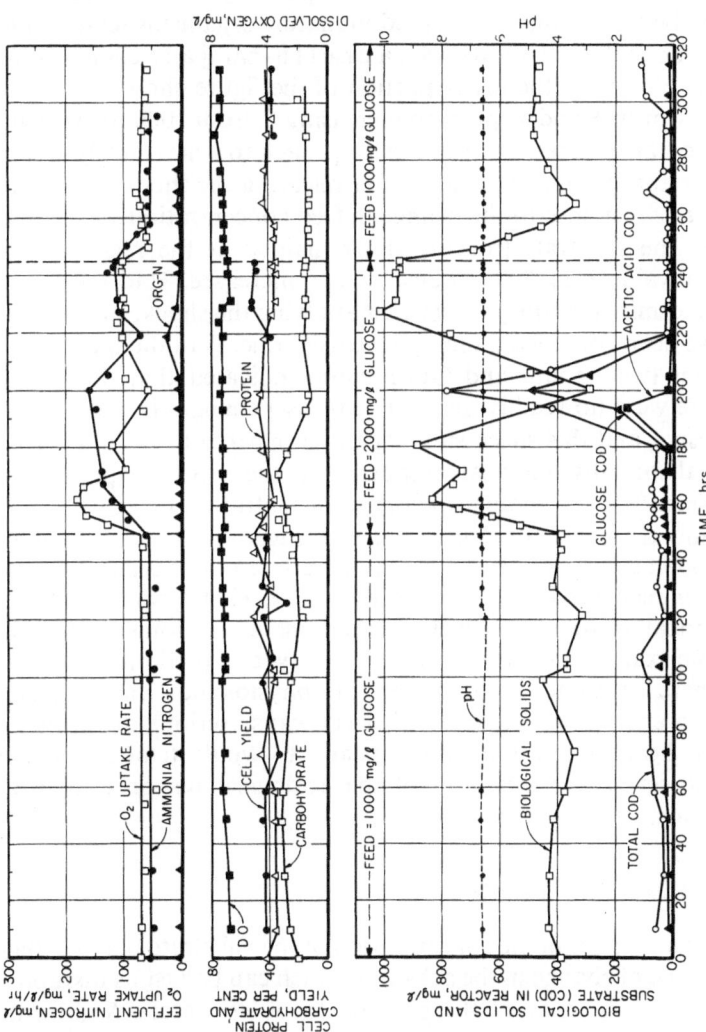

Fig. 11. Primary (metabolic) and secondary (ecological) responses of a mixed microbial population to a doubling of substrate concentration (Thabaraj and Gaudy, 1969)

considerable selective pressure, encouraging changes in predominance of species, and this can lead to serious complications in adjudging success or failure of a system to accommodate an increase in S_i. It would appear that while it may be possible to visualize and analyze the initial response to such a shock as an *en masse* biochemical one, there is reason to consider a secondary (less immediate) but possibly more significant response involving a changeover in predominating organisms set in motion due to the shock. Fig. 11 shows such a case (Thabaraj and Gaudy, 1969). The results plotted in the lower portion of the figure show the "steady state" behavior in S and X prior to increasing S_i from 1000 to 2000 mg/l glucose, the transient response and the approach to a new steady state as well as the return to the initial state after releasing the shock. The immediate *en masse* response to the change in feed concentration was entirely successful. The substrate (total COD) remained at the relatively low steady state value, and X (biological solids) increased. Had the experiment been terminated thirty hours after applying the shock, the response would have been adjudged a highly successful one. Periodic microscopic observations (wet mount and Gram stain) indicated that during this period there were no readily discernible (gross) changes in predominating organisms, and this successful response appeared to be an *en masse* biochemical or physiological response of the existing population. Shortly thereafter, the biological solids concentration decreased, S increased, and there was evidence that a changeover in predominating species was occurring. After a rather severe dilute-out of cells and rise in S, the system recovered and a new steady state level of X was established. The numerical value of cell yield, Y, had increased somewhat and a microscopic examination showed a different type of predominating organisms. There is thus need to consider both physiological and ecological responses to changes in S_i, since either can cause serious disruption of the system. Again, operation with constant X_R would tend to counterbalance the severe consequences of the ecological response, to some extent, at least.

b) Qualitative Shock Loads

Important as changes in the concentration of the substrate may be, there is another type of change in the substrate which can be just as disruptive, or more so, to efficiency of the process. In waste waters, S_i does not represent one carbon source or one type of carbon source. Because the population consists of a mixed microbial community, it might be reasoned that the various substrates present in a waste exert a selective pressure by which various organisms best suited for the metabolism of each substrate develop, thus ensuring the concurrent removal of all carbon sources. Such a view is rather over-simplified, and is not consistent with the operation of such metabolic control mechanisms as catabolite repression and catabolite inhibition. However, even if one assumes it does have some validity, then it must be admitted that if a new substrate appears in the feed stream, the system may not be ecologically poised to accommodate metabolism of the new carbon source in the waste. In such a situation, the total concentration of the carbon source(s), S_i, may not have changed. Thus, rather than a change in quantity of carbon source (i.e., a "quantitative" shock load), the shock may involve a "qualitative" change. Both types of shock may occur simultaneously, further complicating the problems of predictive assessment of response.

Assuming a purely qualitative change, wherein the nature of a portion or all of the total carbon source has been changed, a variety of alternative responses is possible. The microbial population may undergo a period of selection before the new compound is assimilated. On the other hand, the response may not be initiated by a change in predominating species, since those which predominated prior to the change may induce the requisite enzymes to metabolize the new compound(s). However, since the presence of one substrate can cause interference with the metabolism of another, induction of the requisite enzymes for the new compound(s) may be repressed. On the other hand, the new substrate may repress further synthesis of enzyme systems already present prior to the shock, i.e., the newly introduced substrate may be assimilated at the expense of some of those already present. Observation of "diauxie" (diphasic growth) of pure cultures in multi-substrate media formed the basis for Monod's initial concepts regarding genetic control of metabolism (1949).

Phasic growth due to sequential metabolism of substrates has also been shown for mixed microbial populations. Even in systems wherein the population has been pre-selected by growth on one carbon source, utilization of this substrate can be prevented due to preferential use of a newly introduced substrate (Gaudy, 1962). Furthermore, the blockage in removal of one substrate due to the introduction of another is not due solely to the repression of enzyme synthesis. In situations wherein large initial inocula of cells pre-acclimated to one compound are exposed to this compound and another, it is still possible for the newly added compound to block the removal of the substrate to which the population was acclimated (Gaudy, Komolrit, and Bhatla, 1963; Gaudy, Gaudy, and Komolrit, 1963). Such findings led to the conclusion that there was, in addition to repression of enzyme synthesis, a second, more immediately responsive mechanism whereby a compound could block or interfere with the functioning of an existing complement of enzymes in catabolic pathways. The mechanism was envisioned as one acting in accordance with the concept of feedback inhibition which had long been known to exist in biosynthetic pathways (Gaudy, Gaudy, and Komolrit, 1963). Since this mechanism is active in catabolic pathways, it has become known as catabolite inhibition. The enzyme inhibition occurs in an early metabolic step, either at the level of entry or at the first (or early) catabolic reaction(s). Further experimental results and reasoning for postulating the existence of such a mechanism have been reviewed by Gaudy and Gaudy (1966). Paigen and Williams (1970) have recently presented a critical review of this and other metabolic control mechanisms in carbohydrate metabolism.

The significance of the operation of catabolite inhibition in response to shock loadings was first emphasized in batch studies demonstrating the rapidity with which the injection of a new substrate could severely hamper metabolism of a substrate which was actively being metabolized prior to the shock (Komolrit and Gaudy, 1966b). In the most severe cases, complete cessation of removal of the primary substrate can occur.

Manifestations of disruption of substrate removal due to repression and inhibition in response to a change in substrate composition in completely mixed continuous culture reactors are not unlike those observable in batch systems. Providing that the new substrate is not introduced as a massive "slug" directly to the reactor but as a change in the inflowing medium, these systems do exhibit some degree of stability with respect to leakage of carbon source. For example, in a once-through reactor or "chemostat" operating at $D = 0.25 \, \mathrm{hr}^{-1}$ with 1000 mg/l sorbitol as limiting nutrient, changing the feed to 750 mg/l sorbitol plus 250 mg/l glucose or to 250 mg/l sorbitol plus 750 mg/l glucose did not lead to an increase in either compound in the effluent nor to an increase in soluble organic matter (i.e., COD) in the effluent (Komolrit and Gaudy, 1966a). When the feed was changed from 1000 mg/l sorbitol to 1000 mg/l sorbitol plus 1000 mg/l glucose, constituting a combined quantitative and qualitative shock loading, there was again no increased leakage of sorbitol or glucose (periodate- or anthrone-reactive material), but there was a significant increase in COD, indicative of the accumulation of metabolic products in the reaction liquor. Shocks of greater magnitude led to leakage of sorbitol as well as metabolic products.

There are many cultural factors which contribute to the biological response and which complicate attempts to make quantitative predictions of responses in various environmental situations. For example, a combined quantitative and qualitative shock can cause a complete change in the nature of the limiting nutrient. Such a combined shock may not be accompanied by a concomitant increase in the amount of nitrogen source, and the limiting nutrient may change from carbon source to nitrogen source. Such an imbalance can lead to a situation wherein metabolism of the new compound completely replaces metabolism of the former carbon source, which is continuously passed into the effluent (Komolrit and Gaudy, 1966a).

The nature of the limiting nutrient is also related to the magnitude of the effect of dilution rate on the response to qualitative shock loadings. Rather extensive studies have been reported (Grady and Gaudy, 1969) in which the effect of shock loadings of glucose, fructose, or ribose in combination with L-lysine (using a lysine-acclimated mixed microbial population) at dilution rates of 0.167 and 0.083 hr^{-1} have been compared for carbon-limited and magnesium-limited systems. In the L-lysine-glucose system, the degree of repression of lysine utilization due to the presence of glucose was increased at the higher dilution rate in the carbon-limited system, whereas the opposite effect was manifested when magnesium was the growth-limiting nutrient. From these results, as well as other available information in the literature (see Grady and Gaudy, 1969), the explanation for such effects would appear to involve the extent to which each type of condition can permit the accumulation of metabolic intermediates (in the cells or in the medium) since such pools of intermediary metabolic products assumedly contain the co-repressor. In carbon-limited systems, faster growth rates (higher dilution rates) can enhance accumulation of such pools when additional carbon source is supplied because the spread between D and the maximum possible specific growth rate, μ_{max} is less than at lower D's. As μ approaches μ_{max}, the disparity between the rate of dissimilation and assimilation for replication increases and dissimilation products (metabolic intermediates) accumulate, giving rise to higher levels of repressors (or inhibitors). However, in systems not limited by carbon source (e.g., those limited by nitrogen source, magnesium source, etc.), μ cannot increase in response to an increase in carbon source. In this case, μ would be expected to remain equal to D; thus, when an easily degraded carbon source is added, the disparity between the rates of catabolism and anabolism is greater at the lower values of D (i.e., μ), and the size of the metabolic pools is greater, thus increasing the degree of repression (or inhibition).

There are many other factors which add complexity and difficulty to the problem of gaining more definitive understanding of responses to this type of environmental change. Those which have been discussed serve to emphasize some of the basic similarities in results which can be obtained with either pure cultures or mixed microbial populations, but superimposed on all studies using microbial populations selected by the cultural conditions (heterogeneous populations) is the need to be cognizant of the ever-present opportunity and possibility for changes in predominating species which could alter the limits of reliability of predictive formulation of response. It is possible, and it would seem necessary from many bioengineering standpoints, to consider this diversified and ever-shifting microbial community as an "entity", a "biomass", and to attempt to describe its "bio-mechanics" with as much quantitative predictive (mathematical) formulation as possible. For the same reasons (heterogeneous microbial populations) that we must settle for approximate numerical values of the biological "constants" μ_{max}, K_S, and Y, we must also accept an approximate range of limiting boundaries to describe the kinetic mode and extent of accommodation to shock loads.

7. Future Outlook

We have attempted to present some practical kinetic concepts for conventional modes of operation of activated sludge processes and have discussed some mechanisms and kinetic aspects peculiar to processes employing mixed microbial populations for treatment of waste waters. The foregoing presentation was intended not only to help form a basis for more sound conceptual understanding of the process, but to demonstrate how such research can lead to innovations or process modifications which may enhance future progress in this vital area of environmental control. We have limited this discussion essentially to mixed microbial populations in continuous culture and to an activated sludge process which is an application of such continuous cultures to waste purification. This choice was dictated both by space limitation and by a consideration mentioned previously, the belief that greater control of waste treatment processeses is desirable and necessary, and that the completely mixed activated sludge processes would seem to offer the greatest promise of such control at present.

It must be recognized, however, that aerobic fixed bed reactors (trickling filters) have occupied an important place in the field of waste water treatment for some time. Their future utility in the field will, in many respects, depend upon the amount and depth of research effort that knowledgeable investigators are willing to expend upon the study of fixed bed processes. In some respects it may be argued that a fixed bed can function more adequately for the sequential removal of mixed carbon sources in waste waters, because both natural selection and engineering innovation could provide for attachment of specific populations for specific types of carbon sources as they progress through the bed. New developments in material, equipment, and bed construction may serve to open up new avenues for mechanistic biological research. For present-day trickling filters, operational flexibility cannot be achieved by functional plant management. It is not our purpose to downgrade the trickling filter or fixed bed reactor in general for waste water treatment processes. We feel that such processes offer a fertile field for research. Various types of fixed bed reactors and/or fixed beds and fluidized reactors in combination would seem to hold useful possibilities in the field. In recent years there has been a resurgence of research interest in such processes, and more information should be available in the future.

The oxidation pond is another biological waste treatment process which has not been discussed in the present chapter, although the reader should be aware of this method of treatment in which the aim is to maintain aerobiosis for an organotrophic natural population through photosynthetic production of oxygen by a natural algal population. There are instances wherein ample space and low organic loadings combine to make such processes attractive and feasible. There are other instances wherein because of economic considerations the alternative to oxidation ponds is no treatment at all, in which case their use is amply justified. Also, there are times when a first step is logically the installation of oxidation ponds. As loadings increase and space becomes more critical, the mode of treatment can be progressively stepped up, e.g., by addition of mechanical aeration.

A significant problem pertinent to the use of oxidation ponds lies in the fact that in order to produce the oxygen for the aerobic bacterial metabolism of organic matter in a waste water, one must encourage the fixation of carbon dioxide and synthesis of organic matter in the form of algal cells. These are difficult to remove from the pond effluent, and one runs the serious risk of adding as much or more organic matter to the receiving stream as he has caused to be removed from the waste. In addition, recent attention has been turned to the problem of algal blooms in the stream. Continual seeding of the stream with algae, coupled with their possible contribution to bottom sediment and release of the nitrogen and phosphorus they contain are sufficient cause for caution concerning sole reliance on the oxidation pond method of aerobic waste treatment in the future. The concept of using photosynthesis for aeration and employing algae as users of nitrogen and phosphorus as well is indeed a useful one. When it can be coupled with economical separation and perhaps harvesting of the protein produced in the process for its food value (either directly or as feed stock for animal and fish populations), the concept holds some promise in the technologically managed environment of the future.

The concern of biological or biochemical engineering cannot be solely with the on-shore process of waste water purification. Equal attention must be given to the aquatic environment into which the product of the process (the purified effluent) is returned for re-use. The biological engineering and management of this resource is an area of activity of extreme importance which has been relatively unexplored in the biological engineering field. Although some might term this aspect "ecological engineering" or "ecological management", it is really a part of biological engineering which must not be divorced from the "on shore" organic extraction processes. Indeed, the whole bioenvironment, the life support system, must ultimately be viewed and dealt with as one system. Such a unifiying conceptual viewpoint is becoming more and more necessary and, indeed, is being increasingly adopted by forward-looking workers in the field. Concerning the aqueous environment alone, not too long ago water supply, waste water treatment, and stream "sanitation" aspects were considered by many to be separate segments of the field. Today these aspects are considered by most engineers as one system. As the population and attendant industrial and commercial activity increases, the need to consider the fixed water resource as one recycle system requiring technological management becomes more and more apparent. Thus, the behavior and effects of the treatment plant effluent when it re-enters the water resource are as vital a technological concern of the biological engineer as is the treatment process itself.

In recent times many voices have been raised in warning, in protest, and in anger concerning deterioration of the environment, and there is at the same time a worldwide concern and quest for enhancement of the quality of life. Advanced technology is sometimes cited as being a culprit in degrading the environment, yet few would deny the contributions it has made to the quality of life many enjoy today and still more are seeking. Thus, the affluence enjoyed by some people is due largely to advancing technology and the unmitigating drive and inalienable right of all peoples to strive toward such affluence predicts ever-increasing technological sophistication. As the population and its technological (industrial, commercial) activity grow, its natural environment will, of necessity, become one which is technologically controlled by society itself. While the production of food, the generation of power, and the production of the advanced accoutrements of living which man deems necessary to his well-being can exert degradative effects on the life support system, man's intellectual prowess can be directed toward technological control, protection and enhancement of this life support system. Some will abhor the thought that a technologically controlled environment will be (or even could be) the natural environment of the future. Oddly enough, some of these same individuals see population control as the ultimate solution. While control of population growth will and should play an important role in determining the future quality of the environment, it does, after all, represent the most severe form of technological control of the natural ecosystem, and is not an alternative to a technologically-controlled environment but another side of the same coin.

Biochemical engineers have the opportunity and the responsibility of making a unique contribution to environmental management by participating in the necessary task of recycling the fixed water resource, and in the technological management of the aquatic ecosystem. Purified water is the most important and necessary of all "fermentation products", since it is undoubtedly the maintenance of the quality of the fixed water resource which can place an upper limit on the enhancement of the quality of life for all people.

Symbols

a'		ratio of recycle flow to feed flow,
b		ratio of cell mass per unit volume in the recycle flow to cell mass per unit volume in the reactor or reactor effluent,
D	T^{-1}	dilution rate, F/V,
F	$L^3 \cdot T^{-1}$	flow rate, V/t,
K_M		Michaelis-Menten constant, substrate concentration at which velocity of enzyme reaction is 0.5 maximum velocity,
K_S		saturation constant, substrate concentration at which $\mu = 0.5\,\mu_{max}$,
N	L^{-3}	number of cells per unit volume,
S	ML^{-3}	concentration of limiting substrate (in this chapter, S represents the carbon source measured as chemical oxygen demand, i.e., total organic matter),
S_0	ML^{-3}	initial concentration of carbon source,
$\bar{S}$	ML^{-3}	concentration of limiting substrate in the reactor at steady state,
S_i	ML^{-3}	concentration of limiting substrate in the inflowing medium,
t		time,
$\bar{t}$		mean residence time in the reactor,
V	L^3	volume of reaction liquor in the aeration tank,
X	ML^{-3}	mass of cells per unit volume,
$\bar{X}$	ML^{-3}	mass of cells per unit volume in the reactor at steady state,
X_R	ML^{-3}	mass of cells per unit volume in the recycle,
Y		yield coefficient,
μ		specific growth rate, or logarithmic growth rate constant,
μ_{max}		maximum specific growth rate.

References

Busch, A. W., Myrick, N.: Sewage Ind. Wastes **32**, 949 (1960).

Gaudy, A. F., Jr.: Appl. Microbiol. **10**, 265 (1962).

Gaudy, A. F., Jr., Engelbrecht, R. S.: In: Advances in Biological Waste Treatment, p. 11. New York: Pergamon Press 1963.

Gaudy, A. F., Jr., Gaudy, E. T.: Ann. Rev. Microbiol. **20**, 319 (1966).

Gaudy, A. F., Jr., Gaudy, E. T., Komolrit, K.: Appl. Microbiol. **11**, 157 (1963).

Gaudy, A. F., Jr., Goel, K. C., Freedman, A. J.: In: Advances in Water Pollution Research, p. 613. New York: Pergamon Press 1969.

Gaudy, A. F., Jr., Goel, K. C., Gaudy, E. T.: Appl. Microbiol. **16**, 1358 (1968).

Gaudy, A. F., Jr., Goel, K. C., Gaudy, E. T.: Biotech. Bioeng. **11**, 53 (1969).

Gaudy, A. F., Jr., Komolrit, K., Bhatla, M. N.: J. Water Pollution Control Federation **35**, 903 (1963).

Gaudy, A. F., Jr., Ramanathan, M.: Biotech. Bioeng. **13**, 113 (1971).

Gaudy, A. F., Jr., Ramanathan, M., Rao, B. S.: Biotech. Bioeng. **9**, 387 (1967).

Gaudy, A. F., Jr., Ramanathan, M., Yang, P. Y., DeGeare, T. V.: J. Water Pollution Control Federation **42**, 165 (1970).

Gaudy, A. F., Jr., Yang, P. Y., Obayashi, A. W.: J. Water Pollution Control Federation **43**, 40 (1971).

Goel, K. C., Gaudy, A. F., Jr.: Biotech. Bioeng. **11**, 67 (1969a).

Goel, K. C., Gaudy, A. F., Jr.: Biotech. Bioeng. **11**, 79 (1969b).

Grady, C. P. L., Jr., Gaudy, A. F., Jr.: Appl. Microbiol. **18**, 790 (1969).

Herbert, D., Elsworth, R., Telling, R. C.: J. Gen. Microbiol. **14**, 601 (1956).

Herbert, D.: In: Continuous Culture of Micro-organisms, Soc. Chem. Ind. (London), Monograph No. 12, p. 21. New York: The Macmillan Co. 1961.

Komolrit, K., Gaudy, A. F., Jr.: J. Water Pollution Control Federation **38**, 85 (1966a).

Komolrit, K., Gaudy, A. F., Jr.: J. Water Pollution Control Federation **38**, 1259 (1966b).

Komolrit, K., Goel, K. C., Gaudy, A. F., Jr.: J. Water Pollution Control Federation **39**, 251 (1967).

Kountz, R. R., Fourney, C., Jr.: Sewage Ind. Wastes **31**, 819 (1959).

Krishnan, P., Gaudy, A. F., Jr.: Biotech. Bioeng. **7**, 455 (1965).

Krishnan, P., Gaudy, A. F., Jr.: J. Water Pollution Control Federation **40**, R54 (1968).

Monod, J.: Ann. Rev. Microbiol. **3**, 371 (1949).

Monod, J.: Ann. Inst. Pasteur **79**, 390 (1950).

Moser, H.: Carnegie Inst. Wash. Publ. 614 (1958).

Novick, A., Szilard, L.: Science **112**, 715 (1950).

Orwell, G.: Animal Farm. New York: Harcourt, Brace and World, Inc. 1946.

Paigen, K., Williams, B.: In: Advances in Microbial Physiology, Vol. 4, p. 251. New York: Academic Press 1970.

Peil, K. M., Gaudy, A. F., Jr.: Appl. Microbiol. **21**, 253 (1971).

Perret, C. J.: J. Gen. Microbiol. **22**, 589 (1960).

Porges, N., Jasewicz, L., Hoover, S. R.: Appl. Microbiol. **1**, 262 (1953).

Ramanathan, M., Gaudy, A. F., Jr.: Biotech. Bioeng. **11**, 207 (1969).

Ramanathan, M., Gaudy, A. F., Jr.: Biotech. Bioeng. **13**, 125 (1971).

Rao, B. S., Gaudy, A. F., Jr.: J. Water Pollution Control Federation **38**, 794 (1966).

Sawyer, C. N.: In: Biological Treatment of Sewage and Industrial Wastes, Vol. I, p. 3. New York: Reinhold Publishing Corp. 1956.

Schaefer, W.: Ann. Inst. Pasteur **74**, 458 (1948).

Schultze, K. L.: Water Sewage Works **11**, 526 (1964).

Servizi, J. A., Bogan, R. H.: J. Sanit. Eng. Div., Am. Soc. Civil Engrs. **89**, (SA3), 17 (1963).

Siegel, B. V., Clifton, C. E.: J. Bacteriol. **60**, 573 (1950).

Storer, F. F., Gaudy, A. F., Jr.: Environ. Science Technol. **3**, 143 (1969).

Standard Methods for the Examination of Water and Waste Water, 12th edition (1965), American Public Health Association.

Symons, J. M., McKinney, R. E.: Sewage Ind. Wastes **30**, 874 (1958).

Thabaraj, G. J., Gaudy, A. F., Jr.: J. Water Pollution Control Federation **41**, R322 (1969).
Thabaraj, G. J., Gaudy, A. F., Jr.: J. Water Pollution Control Federation **43**, 318 (1971).
Young, T. B., Bruley, D. F., Bungay, H. R., III: Biotech. Bioeng. **12**, 747 (1970).

Professor Anthony F. Gaudy, Jr.
Elizabeth T. Gaudy
School of Civil Engineering and
Dept. of Microbiology
Oklahoma State University
Stillwater OK/USA

CHAPTER 4

Scale-Up of Biological Wastewater Treatment Reactors

W. Wesley Eckenfelder, Jr., Brian L. Goodman, and A. J. Englande

With 22 Figures

Contents

A. General Concepts of Biological Wastewater Treatment

For the purpose of this presentation, the activated sludge process will be considered as that process of wastewater treatment which involves:
The aeration of wastewater for some significant period of time.
The provision of solids-liquid separation at the end of the aeration period.
The discharge of the liquid fraction as process effluent.
The return of some or all of the separated solids to the aeration stage of the process where it is mixed with the influent wastewater.
Other aerobic biological treatment processes include aerated lagoons, stabilization basins and trickling filters. In aerated lagoons only the first step shown above is employed.

1. Conventional Activated Sludge

The Conventional Activated Sludge variation provides for the aeration of settled wastewater, together with sufficient return sludge to produce a mixed liquor solids (MLSS) concentration of from 1200 to 3000 mg/l. The aeration period for domestic wastewater is commonly four to six hours. High strength industrial wastewaters may require aeration periods as long as 24—48 hrs. Activated sludge solids are separated from the treated liquid by quiescent settling. Plug-flow or quasi-plug flow aeration tanks with both the influent wastewater and return sludge (RSSS) being introduced at the influent end are commonly utilized for domestic sewage. Industrial wastewater subject to slugs, spills, vanable plant, etc. usually employ a completely mixed basin. The loading factor of food to micoorganisms (F/M) or ratio of BOD_5 to MLSS is in the range of 0.2 to 0.5 per pound.
Excess activated sludge solids will accumulate in this system. These solids, composed of both biological solids and inerts, accumulate at the rate of about 0.4 to 0.5 pound per pound of BOD_5 removed. Reaeration of the return sludge is sometimes practiced, in which case the solids-accumulation rate will be reduced to some lower level depending on the time of reaeration. If reaeration is employed, the weight of sludge solids or the volume of the reaeration tankage should be considered in computing process loading so that a true comparison with other variants can be made.

2. High-Rate Activated Sludge

The High-Rate variation employs short aeration periods, low activated-sludge concentration levels and, therefore, high process loading. BOD removals are relatively low (40—70%). Sludge return rates are also low, usually about 10% of the average influent flow rate. High solids-accumulation rates are associated with this variation due both to the retention of influent inerts in the system and the lack of sufficient time for significant oxidation of cellular mass formed by the synthesis of removed degradable organics.

3. Step Aeration

The Step Aeration variation utilizes plug-flow aeration tankage with the return sludge being added at the influent end. Settled wastewater is added at several points along the aeration tank. Depending on the actual location of the wastewater addition points, return-sludge aeration may or may not be provided.

Also, depending on the addition point locations, the process might, in essence, really represent Two-Stage Aeration or some other process variant (see subsequent sections). Through the use of stepwise wastewater addition, the volumetric loading is approximately doubled compared to Conventional Contact Stabilization variant.

In the TSA variation, a mixing (contact) time of about three hours is utilized. This period is sufficiently long to provide for adsorption, solubilization, absorption, and the completion of cellular material synthesis. The second aeration stage (reaeration) is also lengthened to some seven hours or more thus providing for the oxidation of the major portion of the synthesized cell mass.

Volumetric loadings applied to TSA systems are the same as those associated with Conventional Activated Sludge. F/M ratios, however, are about half those of the Contact Stabilization variant and approach those of the Extended Aeration variation.

4. Contact Stabilization and Two-Stage Aeration

The two most commonly employed activated sludge process variants utilizing the multi-stage aeration and sludge return flow sheet are contact stabilization and two stage aeration. The difference between these two systems relates to the aeration times employed.

Contact stabilization takes advantages of the phenomenon of rapid sorption of wastewater constituents. Since BOD is rapidly removed from many wastes in this manner by well stabilized activated sludges, the initial process step (the sorption or contact phase) can be of as short duration as 15—30 minutes. The sludge mass is then seperated from the treated liquid and re-aerated (the stabilization phase) for a period of time sufficient to provide for the solubilization of the sorbed organic matter and its subsequent metabolism. The duration of the reaeration phase is commonly 2—6 hours.

5. Extended Aeration

The Extended Aeration variation employs the same flow sheet as Conventional Activated Sludge, but the aeration period is much longer, frequently 24 hrs or more. Loadings are usually in the range of 0.01—0.07 lb. BOD_5/lb. MLSS.

Extended aeration plants are frequently small, almost invariably completely mixed, and return sludge rates are generally high (about equal to the average influent wastewater flow rate). The F/M ratio level is very low, thus, almost complete oxidation of synthesized cell mass results, oxygen uptake rates are low and solids accumulation results mainly from the retention of influent, inert solids in the system.

It should now be clear that a variety of activated sludge process variations are possible. It should be equally clear that these variations are closely related, differing mainly in terms of the mixing regime, flow sheet, and loading level employed. This close relationship is apparent in any study which relates the performance and characteristics of each to their loading level (F/M) range. These facts taken together suggest a useful, rational, activated sludge process description and design scheme.

6. Aerated Lagoons

An aerated lagoon is an earth-dug basin in which biological organisms are allowed to grow and proliferate in an aerobic environment. Oxygen is supplied to the basin by either mechanical or diffused aeration. The turbulence levels created by these aeration systems are high enough to insure uniform distribution of oxygen throughout the basin, but depending upon the type of lagoon system employed, may or may not be sufficient to maintain all of the solids in suspension.

Aerated lagoons are generally classified as either aerobic or facultative, the distinction being based on power level. An aerobic lagoon is a high rate system where soluble BOD is converted to cellular protoplasm with some stabilization occuring. These high-rate lagoons are completely mixed with no solids deposition and must operate at power levels sufficient to maintain all solids in suspension.

In a faculative lagoon the turbulence levels are not sufficient to maintain all of the solids in suspension. Depending on the BOD loading and the suspended solids present in the influent wastewater, some of the solids are deposited in the bottom of the basin where they undergo anaerobic decomposition. The anaerobic by-products are in turn oxidized in the upper aerobic layers of the basin.

7. Design Rational

The solution of any activated-sludge design problem involves the judicious selection of the mixing regime, flow sheet and loading factor to be employed. Thus, the range of common choices can be catalogued as follows:

(A) Mixing Regime.
   1. Plug Flow.
   2. Complete Mixing.
(B) Flow Sheet.
   1. Aeration Only (Aerated lagoons).
   2. Aeration and Sludge Return.
   3. Multi-Stage Aeration and Sludge Return.
   4. Step-Aeration.
(C) Loading Level.
   1. High Rate
      1.0 lb. BOD_5/lb. MLSS or greater.
   2. Conventional
      0.2 to 0.6 lb. BOD_5/lb. MLSS.
   3. Contact Stabilization and Two-Stage Aeration
      0.07 to 0.2 lb. BOD_5/lb. MLSS.
   4. Extended Aeration
      Less than about 0.07 lb. BOD_5/lb. MLSS.

The designer can make a number of combinations within the range of choices presented. The common combination, Conventional Activated Sludge, consists of plug flow, or complete mix, aeration and sludge return, and a loading factor of about 0.3 pound of BOD_5 per pound of mixed liquor suspended solids. This combination is A1 or A2 + B2 + C3.

8. Mixing

Complete mixing cannot be dismissed as just an alternative mixing regime. It must be noted that this mixing mode is a welcome addition to any activated sludge process design. Through the use of the complete mixing regime, oxygen-uptake rate variations due to load fluctuations are minimized by the maintenance of an essentially constant food to microorganism ratio (F/M) [(1—3)]. Thus, Complete Mixing Activated Sludge (CMAS) systems are highly resistant to upset occasioned by shock loadings and the deleterious effects on the activated sludge organisms of low residual oxygen concentration near the influent end of conventional plug flow aeration tans are avoided.

Complete mixing can be defined, therefore, as the maintenance of a uniform solids concentration and oxygen uptake rate level throughout an aeration tank or compartment.

Plug-Flow regimes, on the other hand, feature a progressively decreasing F/M ratio and oxygen-uptake rate from the inlet to the oulet of an aeration tank. Strict plug-flow conditions are not often observed and most so-called plug-flow aeration tanks are more properly referred to as semi-plug-flow.

Step aeration or the introduction of wastes at several points along an otherwise plug-flow tank tends to equalize the oxygen uptake rates observed from inlet to outlet. At the same time, the concentration of activated-sludge solids decrease from inlet to outlet which facilitates solids-liquid separation during final clarification. The zone-settling rate of activated sludge solids is a function of their concentration, among other things. Over a relatively wide range of solids concentrations, the higher the concentration, the lower the settling rate. Thus, there is a distinct advantage inherent in the ability to minimize, to a degree, the solids concentration entering the final clarifier.

9. Flow Sheet

Typical flow schemes are illustrated in Fig. 1. The Aeration Only flow sheet in the past has been underutilized, but has great potential utility especially in the pretreatment of industrial and commercial wastes. Through the utilization of this simple flow sheet, a biodegradable industrial waste having a BOD_5 concentration of 1000 mg/l can be reduced to a 330 mg/l BOD_5 concentration by employing a three-day detention or 265 mg/l with a five-day detention time. The use of the Aeration Only flow sheet eliminates the need for a clarifier and, thus, any possible clarification problems. Also, it eliminates sludge-return pumping and piping. Thus, both the process and equipment are simplified, operational attention requirements are reduced and energy input requirements are minimized. The aerated lagoon is a typical process employing the Aeration Only flow scheme. The waste stabilization pond is also described by

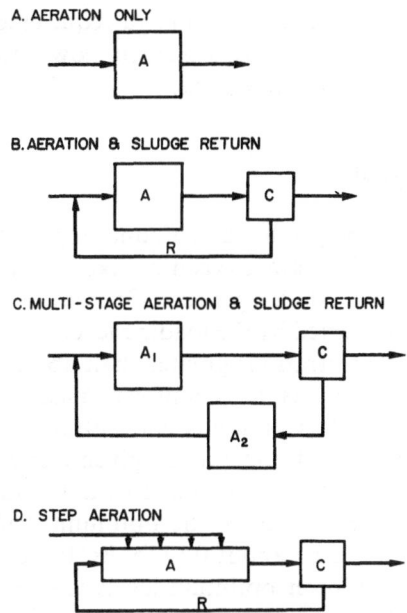

Fig. 1. Activated sludge flow sheet

Aeration Only. However, no artificial aeration or mixing is used. Substrate reduction and oxygen production is the result of a symbiotical relationship between algae and bacteria and mixing is induced by wind. For a more comprehensive discussion of these two treatment processes, the interested reader should consult Eckenfelder [4].

The Aeration and Sludge Return flow sheet is the most commonly employed activated-sludge process variant. The utilization of sludge return permits the maintenance of much higher sludge-age values than can be practically achieved with the Aeration Only flow sheet. Much lower biological sludge accumulation rates result from these higher sludge ages and effluent solids levels are minimized through the use of a final clarifier.

A properly designed treatment plant utilizing the Aeration and Sludge Return flow sheet would be capable of reducing the 1000 mg/l BOD_5 concentration of the waste in the preceding example to about 25 mg/l in a matter of a few hours.

Multi-Stage Aeration and Sludge Return flow sheets are commonly employed when reaeration of the return sludge is desirable. In both the Contact Stabilization and Two-Stage Aeration variants of the activated-sludge process, the use of the Multi-Stage Aeration flow sheet potentially permits a reduction in aeration tankage requirements as compared to that of the Aeration and Sludge Return flow sheet to achieve equal overall results on a comparable basis.

10. Loading Level

The loading level employed has a profound impact on process efficiency, biological accumulation rate, oxygen uptake rate and other prime parameters of the activated-sludge process. One of the most important characteristics which is affected by the loading level is the sludge settling rate. At sludge-solids concentrations greater than about 1000 mg/l, activated sludge settles, more or less, as a uniform mass (displays zone-settling characteristics). The effect of solids concentration on settling rate has already been noted. In addition, for any given waste an optimum settling rate can be demonstrated as associated with a loading level specific for that waste. For domestic wastes, the optimum loading with respect to settling rate appears to be about 0.17 lb. BOD_5/lb. activated sludge solids. Both higher and lower optimum levels are commonly encountered with other types of wastes.

At very low loadings, the activated-sludge flocs deflocculate or disintegrate with a resultant loss of high concentration of fine particles in the plant effluent. At very high loadings, filamentous growth and dispersed growth occurs, settling rate is drastically decreased, and high concentrations of solids also appear in the plant effluent.

The selection of the proper loading level is then an optimization problem, the solution of which must take into consideration not only setting rate, but also solids accumulation, oxygen uptake rate, and BOD removal.

B. Mathematical Models and Design Example

1. General Mathematical Formulations for Design

A number of activated sludge process math models have been advanced which present relationships useful in data correlation and process design. Notable among these are those of Eckenfelder [4, 5] and McKinney [1, 2]. In addition, Goodman [6, 7] has presented a variety of process design examples covering the commonly employed range of activated sludge process variants.

Table 1. McKinney math model [1, 2, 6]

$$F = \frac{F_i}{K_m t + 1} \tag{1}$$

$$M_a = \frac{K_s F}{\dfrac{1}{t_s} + K_e} \tag{2}$$

$$M_e = 0.24 \, K_e \, M_a \, t_s \tag{3}$$

$$M_i = f_i \, SS_{inf} \frac{t_s}{t} + 0.1 \, (M_a + M_e) \tag{4}$$

$$M_T = M_a + M_e + M_i \tag{5}$$

$$M_{T_v} = M_a + M_e + M_{iinf} \tag{6}$$

$$BOD_{5_e} = F + K \, M_{T_e} \cdot \frac{M_a}{M_T} \tag{7}$$

$$O_u = \frac{a''(F_i - F)}{t} - \frac{b''(M_a + M_e)}{t_s} \tag{8}$$

$$x = \frac{\dfrac{0.77 \, a \, S_r}{t}}{X_v \left(\dfrac{1}{t_s} + K_e\right)} \tag{9}$$

Table 2. Eckenfelder math model [4, 5]

$$S_e = \frac{S_0}{k \, X_v t + 1} \tag{1}$$

$$\Delta X_v = \frac{a \, S_r}{t} - b \, X_v \tag{2}$$

$$\Delta X_v = \frac{a \, S_r}{t} - k_b x \, X_v$$

$$\Delta X_v = \frac{a \, S_r}{t} + f X_0 - b \, X_v$$

$$x = \frac{a \, S_r + k_b X_v - \sqrt{(a \, S_r + k_b X_v)^2 (4 \, k_b X_v)(0.77 \, a \, S_r)}}{2 \, k_b X_v} \tag{3}$$

$$x = \frac{a \, S_r + k_b X_v + f X_0 - \sqrt{(a \, S_r + k_b X_v + f X_0)^2 (4 \, k_b X_v)(0.77 \, a \, S_r)}}{2 \, k_b X_v}$$

$$BOD_{5_e} = S_e + f X_{v_e} \tag{4}$$

$$R_r = \frac{a' S_r}{t} + b'' k_b x \, X_v \tag{5}$$

$$R_r = \frac{a' S_r}{t} + b' X_v$$

$$\frac{F}{M} = \frac{24 S_0}{X_v t} \tag{6}$$

Table 3. Nomenclature and relationships

t	aeration period, days or hours
S_0	$F_i =$ influent BOD_5, mg/l
S_e	$F =$ effluent soluble BOD_5, mg/l
S_r	$F_i - F =$ removed BOD_r, mg/l
X_0	$f_i SS_{inf} =$ inert influent suspended solids, mg/l
X_v	$M_a + M_e$ (in the case of soluble wastes) $=$ volatile suspended solids, mg/l
X_{v_e}	$M_{T_{v_e}} =$ effluent volatile suspended solids, mg/l
M_a	active mass (as volatile suspended solids), mg/l
M_e	endogenous mass (as volatile suspended solids), mg/l
M_i	inert mass, mg/l
M_{iinf}	influent inert volatile suspended solids, mg/l
M_T	total mass, mg/l
G	$t_s =$ sludge age, days
BOD_{5_e}	effluent BOD_5 (total), mg/l
R_r	$O_u =$ oxygen uptake rate, mg/l/day or mg/l/hour
F/M	food to microorganism or food to mass ratio, lb/lb/day
k	$\dfrac{K_m}{X_v} =$ BOD removal rate, l/mg-day
k	overall BOD removal rate, day^{-1}; $K = kXv$
K_s	$aK_m =$ synthesis rate, day^{-1}
k_b	$K_e = \dfrac{b}{x} =$ endogenous rate, day^{-1}
a	mass yield rate, 1b/1b BOD removed
x	degradable fraction, % (as decimal fraction)
a'	$1 - a =$ synthesis oxygen demand rate, lb/lb
b'	$b'' b = b'' k_b x =$ endogenous respiration oxygen demand rate, lb/lb
b''	oxygen equivalent of the degradable volatile suspended solids, lb/lb
a''	$\dfrac{BOD_u}{BOD_s} =$ ultimate oxygen equivalent of the substrate removed, lb/lb
$\triangle X_v$	solids accumulation rate, mg/l/day

Recent activated sludge wastewater treatment designs have featured one or more completely mixed aeration basins. This mixing mode offers the advantage of dispersing and mixing the incoming wastes throughout the basin contents, thus dampening fluctuations in influent constituents and enhancing process stability. Because of the utility and greatly increased use of this mixing mode, it will be considered in some detail here. The reader is referred to the references previously cited for discussions of other mixing modes and process variants not considered here.

The process math models of Eckenfelder and McKinney are presented in Tables 1 and 2. The appropriate nomenclature is given in Table 3. In order to illustrate the utilization of these models in data correlation and process design, the following example will be employed.

2. Example Problem

An organic chemicals plant is considering alternative aerobic treatment processes for disposal of their wastewaters. The present regulatory agency requirement specifies an average effluent BOD_5 of 30 mg/l. The prototype wastewater characteristics are:

Flow: 0.75 mgd.
BOD_5: 650 mg/l (after 24 hour equalization).
SS: NONE.

Table 4. Pilot plant data (aeration with sludge return)

t day	S_0 mg/l	S_e mg/l	S_r mg/l	X_v mg/l	G day	ΔX_v mg/l /day	F/M lb/lb	SDI	X_e mg/l	x
2.17	650	5	647	3000	66	45	0.1	0.9	9	0.19
1.08	650	10	642	3000	28	107	0.2	3.0	10	0.35
0.54	650	19	631	3000	9.7	309	0.4	2.8	12	0.54
0.363	650	29	622	3000	5.1	588	0.6	1.25	20	0.62
0.27	650	40	610	3000	4.1	731	0.8	0.65	35	0.66
0.217	650	48	602	3000	2.6	1154	1.0	0.4	50	0.69
0.108	650	90	560	3000	1.2	2500	2.0	0.1	80	0.74

Table 4 presents data obtained by a complete mix activated sludge pilot plant. The design coefficients $a=0.55$, $k_b=0.20$, and $a''=1.23$ were derived from the pilot plant's operational data. The coefficient, b'', is usually considered a constant of 1.42. Predictions of prototype substrate removal, sludge accumulation, and oxygen requirements are illustrated in the following calculations:

2.1. BOD Removal

BOD_5 removal data is presented in Figs. 2, 3, and 4. It can be seen from Fig. 2 that in order to achieve the desired effluent BOD_5 concentration, an aeration period, t, of about 0.36 day or greater must be employed.

$$t = \frac{S_r}{k X_v S_e} \tag{1}$$

$$t = \frac{620}{0.02\,(3000)\,30} = 0.344 \text{ day}.$$

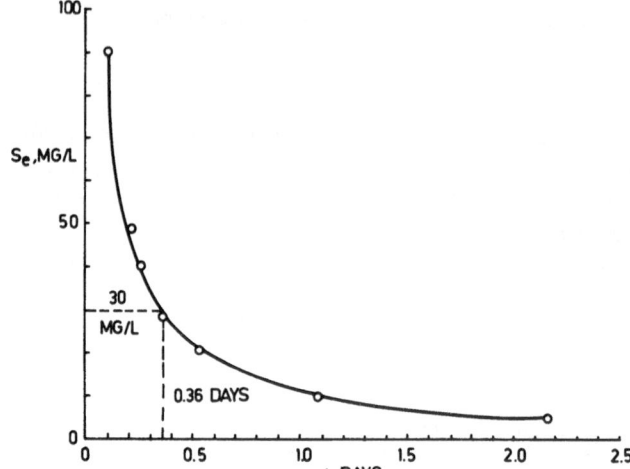

Fig. 2. BOD removal pilot plant data

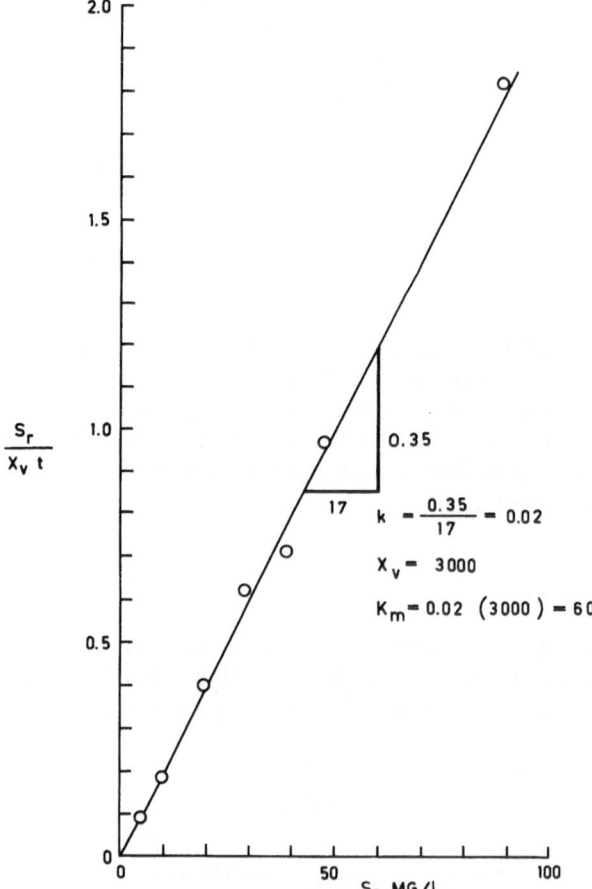

Fig. 3. Determination of BOD removal rate coefficient pilot plant study

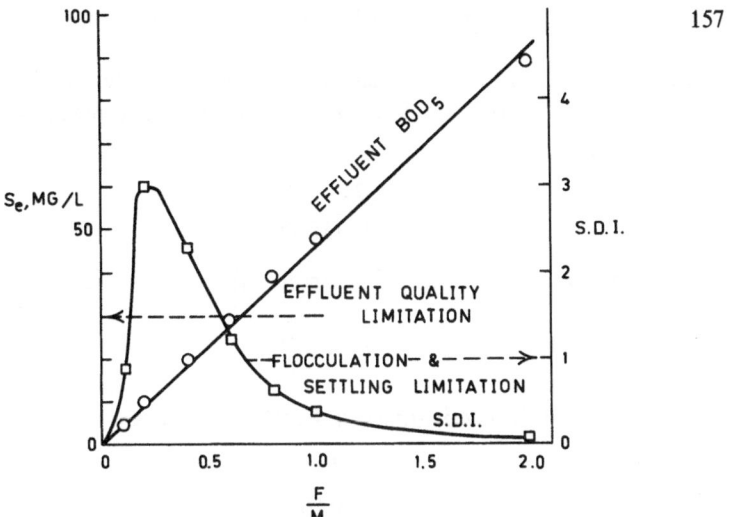

Fig. 4. Effect of loading level on BOD_5 removal and sludge density index

The BOD removal data is correlated, as shown in Fig. 3, according to the relationship:

$$k\,S_e = \frac{S_r}{X_v t} \tag{2}$$

$$k = \frac{620}{3000\,(0.344)\,30} = 0.02 \text{ day}^{-1}$$

But, the total effluent BOD_5 is the sum of the soluble effluent BOD_5, S_e, and the five day oxygen equivalent of the degradable effluent suspended solids.

$$BOD_{5_e} = S_e + 5\,(b''\,k_b\,x\,X_{v_e})$$

$$BOD_{5_e} = 30 + 5\,(1.42)\,(0.2)\,(0.62)\,(20) \tag{3}$$

$$= 30 + 17.6 = 47.6 \text{ mg/l}.$$

Thus, an aeration period of about 0.54 day will be required to achieve the desired effluent quality.

$$BOD_{5_e} = 19 + 5\,(1.42)\,(0.2)\,(0.54)\,(12)$$

$$= 19 + 9 = 28 \text{ mg/l}.$$

If, however, sludge bulking is to be avoided, the F/M value must be maintained at or below some limiting value and the aeration period cannot be less than some corresponding limiting value, see Fig. 4.

$$t = \frac{24\,S_0}{(F/M)\,X_v} \tag{4}$$

$$t = \frac{24\,(650)}{0.66\,(3000)} = 7.9 \text{ hours} = 0.33 \text{ day}.$$

Thus, in the present case, the design aeration period will be based on the rate of BOD_5 removal and the F/M value actually maintained in the treatment system will be:

$$\frac{F}{M} = \frac{24\,(650)}{13\,(3000)} = 0.40.$$

BOD_5 removal data can also be correlated by the relationship:

$$k_m = \frac{S_r}{t\,S_e}. \tag{5}$$

Thus, the unmetabolized substrate BOD_5 in the plant effluent would be:

$$S_e = \frac{631}{0.54\,(60)} = 19 \text{ mg/l}.$$

Or, based on the influent BOD_5, the soluble effluent BOD_5 can be computed as:

$$F = \frac{F_i}{K_m t + 1} \tag{6}$$

$$F = \frac{650}{60\,(0.54) + 1} = 19 \text{ mg/l}.$$

2.2. Solids Accumulation

Solids will accumulate in the treatment system with time due to the synthesis of removed BOD into protoplasm. This gross synthesis value is partially offset by the subsequent oxidation of some amount of the synthesized cellular material. Thus, for soluble wastes, the volatile suspended solids in the treatment system will consist of living cells and the organic, non-degradable, residues of cell oxidation.

Thus, for soluble wastes, the volatile solids in the treatment system at a given sludge age will be:

$$X_v = \frac{\dfrac{a\,S_r}{t}}{\dfrac{1}{G} + x\,k_b}, \tag{7}$$

$$X_v = \frac{\dfrac{0.55\,(629)}{0.54}}{\dfrac{1}{9.7} + 0.54\,(0.2)} = 3035 \text{ mg/l}.$$

This value can also be determined in several other ways:

$$\Delta X_v = \frac{a \, S_r}{t} - k_b \times X_v \, , \tag{8}$$

$$\Delta X_v = \frac{0.55 \, (629)}{0.54} - 0.2 \, (0.54) \, 3035 \, ,$$

$$\Delta X_v = 640.65 - 328 = 312.65 \; \text{mg/l/day} \, ,$$

$$\Delta X_v G = X_v \tag{9}$$

$$X_v = 312.65 \, (9.7) = 3033 \; \text{mg/l/day} \, ,$$

also

$$M_a = \frac{K_s F}{\dfrac{1}{ts} + K_e} \, , \tag{10}$$

$$M_a = \frac{0.55 \, (60) \, 19}{\dfrac{1}{9.7} + 0.2} = 2069 \; \text{mg/l} \, ,$$

$$M_e = 0.24 \, (K_e) \, (M_a) \, (t_s) \, , \tag{11}$$

$$M_e = 0.24 \, (0.2) \, (2069) \, (9.7) = 963 \; \text{mg/l} \, ,$$

$$M_{T_v} = M_a + M_e, \; \text{soluble wastes} \, , \tag{12}$$

$$M_{T_v} = 2069 + 963 = 3032 \; \text{mg/l} \, .$$

In the case of wastes containing inert solids, these will also be accumulated in the treatment system.

$$\Delta X_v = \frac{a \, S_r}{t} + f X_0 - b X_v \, . \tag{13}$$

2.3. Oxygen Requirements

Oxygen will be required for both synthesis and oxidation of cellular material.

$$R_r = \frac{a' \, S_r}{t} + b'' k_b \times X_v \, , \tag{14}$$

$$R_r = \frac{0.45 \, (629)}{0.54} + 1.42 \, (0.2) \, (0.54) \, (3000)$$

$$= 524 + 460 = 984 \; \text{mg/l/day} = 41 \; \text{mg/l/hour} \, ,$$

or

$$O_u = \frac{a''\,(F_i - F)}{t} - \frac{b''\,(M_a + M_e)}{t_s}, \tag{15}$$

$$O_u = \frac{1.23\,(629)}{0.54} - \frac{1.42\,(3000)}{9.7}$$

$$= 1433 - 439 = 994 \text{ mg/l/day} = 41 \text{ mg/l/hour}.$$

BOD Removal Kinetics

The removal of BOD from a wastewater in a lagoon can be expressed by the relationship:

$$S_e/S_0 = \frac{1}{1 + Kt}. \tag{16}$$

In this equation, the overall reaction rate coefficient, K, includes the effects of biological solids and is generally used for convenience. The specific reaction rate coefficient, k, is independent of the microorganism concentration and is a function of the nature of the waste. Thus, equation can be expressed as:

$$S_e/S_0 = 1/1 + kX_vt. \tag{17}$$

Eq. (17) implies that the aeration volatile suspended solids, X_v, are proportional to the active biomass in the process. It has been shown that the active fraction of volatile solids in a fully mixed aeration basin decreases with retention time. This decrease in active mass is attributed to the accumulation of inert mass through extended endogenous metabolism. Eq. (18) can be related to active mass:

$$S_e/S_0 = \frac{1}{1 + k \times X_v t}. \tag{18}$$

The fraction of active mass may be expected to vary from approximately 0.75 for detention times of less than 2 days to 0.5 for detention times of six days. In a fully mixed, flow-through lagoon, the equilibrium biological solids concentration can be defined as:

$$x X_v = \frac{a s_r}{1 + bxt} \tag{19}$$

Combining Eqs. (19) and (20) defines the soluble effluent BOD as:

$$S_e = \frac{1 + bxt}{akt}. \tag{20}$$

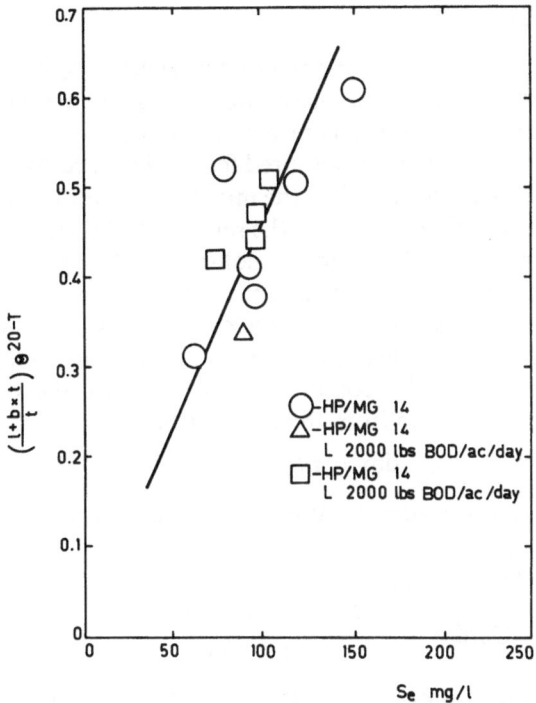

Fig. 5. BOD removal as a function of effluent concentration

Results from a number of lagoons treating pulp- and paper-mill effluents
are shown in Fig. 5.

At low power levels, a portion of the suspended solids may deposit on
the bottom of the basin. The degree of deposition at any power level
will depend on the concentration of BOD and influent suspended solids
in the wastewater. At low BOD and suspended solids concentrations,
the biological growth is dispersed and tends to remain in suspension
even at very low power levels. It has been shown that, for pulp and paper
mill wastewaters at organic loadings less than 2000 lbs BOD/acre/day,
the solids are dispersed at power levels as low as 4 HP/MG and Eq. (20)
may be expected to apply.

In aerated lagoons, depending on mixing intensities, baffling, etc., the
effluent suspended solids may exceed 50 mg/l and will contribute to the
total effluent BOD depending on the active biomass content. The total
effluent BOD can be estimated by:

$$BOD_5 \text{ (effluent)} = S_e + 0.84 \times X_v. \tag{21}$$

Oxygen Requirements

The total aeration horsepower for any pond is determined by the oxygen required for the oxidation of the incoming degradable matter plus that oxygen needed for the endogenous respiration of the biomass produced in the process. If the horsepower required for mixing exceeds the oxygen input needed for BOD removal, then the mixing level controls the design. Due to the higher rate of degradation, the release of BOD from deposited solids in the summer, and the lower solubility of oxygen of higher temperatures, the aeration requirements must be designed for the conditions in the summer months. Therefore, the oxygen requirements are usually controlled wither by summer operation conditions or mixing requirements. In the completely mixed basin, the oxygen requirements can be estimated from the relationship:

$$O_2/day = a' S_r + b' x X_v \qquad (22)$$

where $a' =$ oxygen consumed per unit substrate removed, $b' =$ endogenous rate coefficient (day^{-1}).
Combining Eqs. (19) and (22):

$$O_2/day = \frac{S_r[a'(1 + bxt) + 1.4\,ab]}{1 + bxt} . \qquad (23)$$

This gives the pounds of O_2 required per day. Occasionally it becomes necessary to modify oxygen requirements for anaerobic feedback from benthal deposits.
The oxygenation transfer is given in lb $\cdot O_2/HP - hr$ by:

$$N = N_0 ((C_{sw} - C_L)/C_{Lo}) \,\alpha\, 1.02^{(T - 20)} \qquad (24)$$

where

N_0 = oxygen transfer at 20° C, zero D. O. (lbs $O_2/HP - hr$),
T = temperature (°C),
C_s = saturation oxygen concentration depending on the temperature and elevation (mg/l),
C_L = oxygen concentration in pond (mg/l),
C_{Lo} = saturation oxygen concentration at 20° C (mg/l),
α = ratio of oxygen transfer rate in waste to that in water.
The value of N_0 depends on the nature of the aeration equipment and on the power level in the aeration basin.

Temperature Effects

The performance of aerated lagoons is significantly influenced by changes in basin temperature. The basin temperature in turn is influenced by the temperature of the influent wastewater and the ambient air temperature. A thermal balance of aeration basins results in the relationship:

$$t/D = \frac{(T_i - T_w)}{f(T_w - T_a)} \qquad (25)$$

where

D = basin depth,
T_i = influent waste temperature (°F),
T_w = pond temperature (°F),
T_a = ambient air temperature (°F),
f = proportionality coefficient.

The temperature effect on the biological reaction rate is estimated by a modification of the van't Hoff-Arrenhenius Equation:

$$k_T = k_{20}\, \Theta^{T-20} \qquad (26)$$

where Θ = temperature coefficient.

C. Evaluation of Laboratory Bench Scale and Pilot Plant Data for Progress Design

Due to the complex and variable nature of industrial wastewaters, it is necessary in many, if not most, instances to conduct laboratory studies to obtain pertinent parameters and coefficients for prototype biological treatment plant design. The scale and magnitude of such studies will be dictated by the treatment schemes applicable, the nature of the pollutants, and the funding available. These studies should be geared to the establishment of the most effective and economical treatment system and to proper facility sizing. The reproduction of optimal laboratory conditions to pilot or full scale operation is defined as "scale-up". The methodology, applicability, and limitations of laboratory and pilot plant studies for prototype design are reported herein.

1. Methodology

It is necessary that a general format be adhered to if scale-up is to be successful. First, representative wastewater samples must be collected and their relevant properties characterized. Screening analyses are next conducted for toxicity and other deleterious qualities of the wastewater under consideration. Treatability studies are then performed to determine the most economical treatment scheme, the coefficients to be used for prototype design, and operational problems which might be expected. Ford [8] and Eckenfelder and Ford [5] have treated this subject thoroughly and this section contains a synopsis of their findings.

1.1. Wastewater Sampling and Characterization

"Representativeness" of wastewater samples used in model tests is essential if prototype performance is to be simulated. The first step, then, in any treatability program is a pollution survey to provide sample information with regard to flow, variability, and general characteristics of all waste streams to be treated. The frequency and type of sampling (whether grab, composite, or continuous) will be dictated by the nature of the process under consideration and the tests to be conducted. For example, batch processes should be composited over the duration of each discharge and weighted according to flow; continuous processes should be composited and weighted according to flow. The compositing schedule will depend upon the variability in waste characteristic with respect to proposed treatment facility operation or effect on the receiving water (see Table 5).

Table 5. Suggested sampling or compositing schedule

Characteristic	High variability	Low variability
BOD[a]	4 hour	12 hour
COD or TOC[a]	2 hour	8 hour
Suspended solids	8 hour	24 hour
Alkalinity or acidity	1 hour grab	8 hour grab
pH	Continuous	4 hour grab
Nitrogen and Phosphorus[b]	24 hour	24 hour
Heavy metals	4 hour	24 hour
Temperature	2 hour	8 hour

[a] The compositing schedule where continuous samplers are not used depends on variability, i.e., 15 min for high variability to 1 hour for low variability.
[b] Does not apply to nitrogen or phosphorus wastes (e.g. fertilizer).

The duration of the sampling program must also be defined. For wastes of relatively uniform composition (e.g., those from the paperboard industry) several days should be sufficient to establish design criteria. For plants with a variety of products and production schedules, such as a diverse chemical plant, the survey should be programmed to cover all major production schedules.

After samples have been collected, they must be characterized to determine properties related to treatability. While no common format exists, a basic schedule necessary for defining most wastewaters is outlined in Table 6.

It should be noted that preservation of samples is required prior to chemical analysis. Biological activity can be minimized by cooling the sample to 4° C. Acidification is also used to preserve the organic and

Table 6. Characterization schedule for raw wastewaters

Test	Analytical procedure	Purpose
A. Suspended Fraction		
Suspended Solids (SS)	*Standard Methods*	determine primary clarification requirements
Volatile Suspended Solids (VSS) COD, BOD of suspended fraction	*Standard Methods*	determine organic oxygen demand of suspended fraction
BOD, COD – Total; BOD, COD-Filtered Chemical analysis of suspended fraction		determine potentially toxic substances associated with suspended fraction
Grease, sand, grit, etc.	*Standard Methods*	
B. Total Sample		
COD, BOD, TOD	*Standard Methods* automated techniques	organic strength of wastewater in terms of oxygen demand
TOC	automated techniques	organic strength in terms of organic carbon
pH, acidity, alkalinity	*Standard Methods*	neutralization requirements
Total dissolved solids	*Standard Methods*	effect on unit processes, effluent quality criteria
Ammonia and nitrate nitrogen	*Standard Methods*	nutrient availability
Total phosphate	*Standard Methods*	nutrient availability
Oils	*Standard Methods*	effect on unit processes; need for pretreatment
Specific analysis as required (phenols, heavy metals, etc.)	*Standard Methods*	potential toxicity definition
Chlorides	*Standard Methods*	potential toxicity; effluent quality requirement
C. Miscellaneous Analyses		
Flow, and patterns of flow	weirs, flumes, etc.	sizing unit processes
Temperature	in site	heat balance
Color, toxicity	*Standard Methods*	effect on unit processes, effluent quality criteria

inorganic constituents. The technique employed will depend on the volume of sample, the nature of the wastewater, and the required analytical tests.

Upon completion of the sampling and characterization phase, the screening and treatability phases can commence.

1.2. Screening Analysis for Biological Treatability

It is often necessary in the course of conducting treatability studies to analyze each of several streams within an industrial complex either for absolute toxicity or for substances which may exert a deleterious effect on biological treatment systems. The necessary screening tests can be conducted using either manometric techniques or series of batch reactors.

The Warburg Respirometer can be used to obtain a cursory estimate of the effect of various waste streams on biological cultures manometrically by virtue of the biological response in terms of oxygen uptake for different classifications and concentrations of wastewaters. The generalized use of the Warburg Respirometer, shown in Fig. 6 has been described

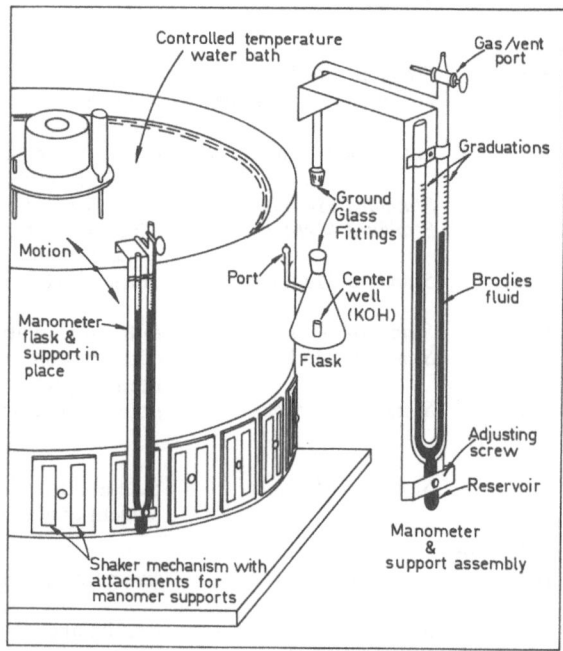

Fig. 6. Warburg respirometer

previously [9]. For purposes of screening, an acclimated biological seed
and the wastewater in question are added to the manometric flasks. As
the microorganisms utilize the organic constituents of the wastewater,
the corresponding pressure differential in the closed system caused by
oxygen utilization is recorded. The cumulative utilization values can be
plotted for various reaction times, and the waste stream or test concentration which shows low oxygen utilization is noted as the toxic or
inhibitory threshold (allowances should be made for lag periods of oxygen utilization caused by seed acclimation requirements). This is graphically illustrated in Fig. 7, which indicates that a specific waste stream is
toxic or inhibitory when the concentration of the particular waste exceeds 10% by volume [10].

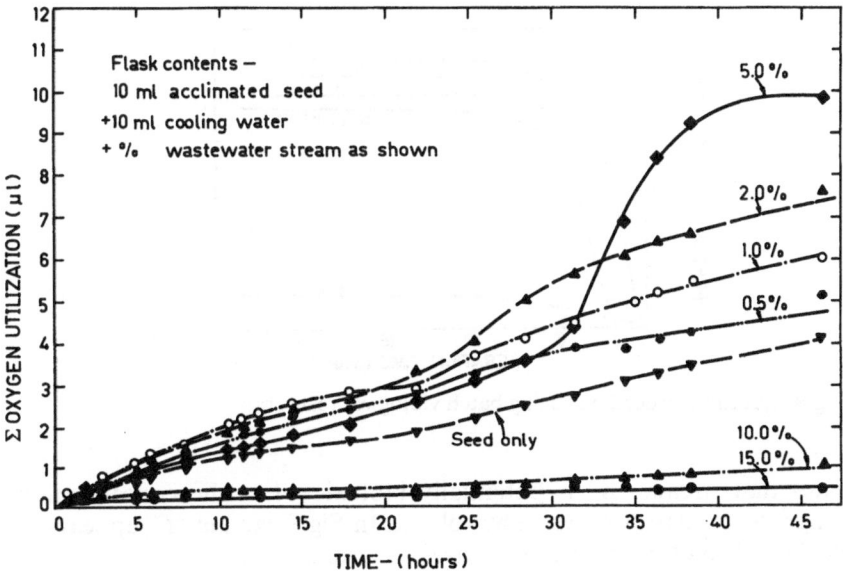

Fig. 7. Warburg applications

A series of batch biological reactors (bench scale) can be used to accomplish essentially the same purpose as the Warburg or manometric approach. An acclimated seed is added to each of a series of batch reactors.
Various concentrations of a single waste stream or wastewater samples
from different streams are then added to each reactor at a volume pro-
portioned by flow. The mixed contents are aerated for 2 to 3 days and
apparent toxicity or inhibition is indicated using COD or BOD removal
as the comparative index, as illustrated in Figs. 8 and 9 [5]. Fig. 8 indi-

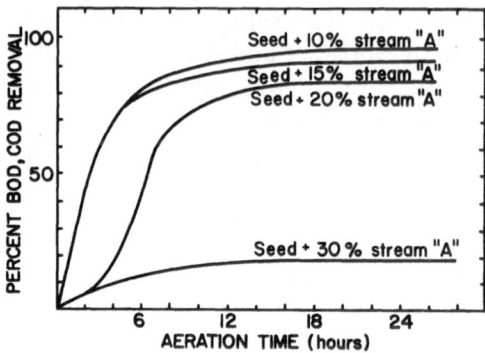

Fig. 8. Screening procedures using batch biological reactors

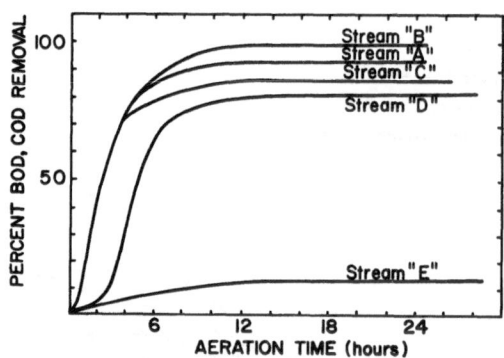

Fig. 9. Screening procedures using batch viological reactors

cates that stream "*A*" is potentially biotoxic when the concentration ranges from 20 to 30% waste by volume. In Fig. 9, stream "*E*" appears to be the only biotoxic waste of the streams tested.

1.3. Treatability

Acclimation of biological seed is in many cases necessary with industrial wastes before design data can be collected. Although continuous or batch units similar to those shown in Figs. 10 and 11 may be used, acclimation is best achieved in continuous reactors. This approach most nearly characterizes the ecosystem which will predominate within the prototype. Once substrate removal stabilizes, acclimation can be assumed. The time required will vary depending on the complexity of the wastewater.

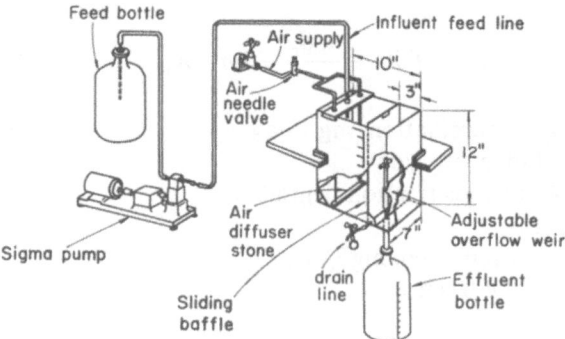

Fig. 10. Actvated sludge flow diagram

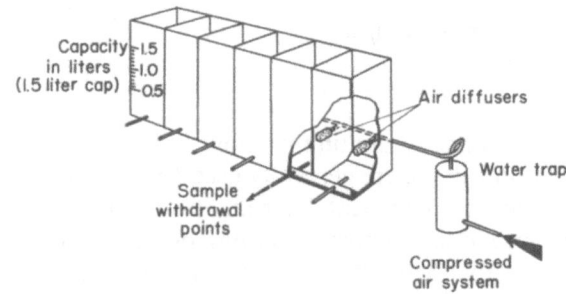

Fig. 11. Batch reactor

Common wastewater treatment plant alternatives include activated sludge, extended aeration, aerated lagoons, and waste stabilization ponds. Procedures for evaluating the performance of each of these alternatives are outlined below.

a) *Activated Sludge.* The previously developed complete mix, mathematical models describing kinetic substrate removal rate (first-order approximation), solids production, and oxygen requirements are summarized as follows:

1. Substrate Removal

$$\frac{S_0 - S_e}{X_v t} = k S_e \qquad (27)$$

S_0 = applied BOD, COD, mg/l,
S_e = effluent BOD, COD, mg/l,
X_v = MLVSS, mg/l,
t = detention time, days,
k = reaction rate coefficient, days^{-1}.

2. Sludge Production

$$X_v = f\,X_0 + \frac{a\,S_r}{t} - kb\,x\,X_v \qquad (28)$$

X_v = net sludge accumulation, mg/l/day,
X_0 = influent sludge, mg/l/day,
f = 1 – influent solids fraction degraded,
a, k_b = coefficients,
S_r = BOD, COD removed, mg/l/day,
x = biodegradable fraction.

3. Oxygen Utilization

$$R_r = \frac{a'\,S_r}{t} + b'\,X_v \qquad (29)$$

R_r = oxygen required, mg/l/day,
a', b' = coefficients.

Bench, eight-liter capacity continuous-flow units, similar to the schematic illustrated in Fig. 10, are generally used to develop the coefficients necessary for scale-up. Coefficient determination procedure entails operating three or four model reactors at varying organic loading rates (from approximately 0.2 to 1.5 lbs BOD_5/day/lb MLVSS) until a quasisteady state condition is obtained, and subsequently measuring parameters of interest. The following analytical and sampling schedule should be considered for each organic loading:

Analysis	Frequency	Raw[a] waste	Mixed[b] liquor	Effluent[c]
COD, BOD, or organic carbon, mg/l (filtered and unfiltered composite samples)	3/week	×		×
pH	daily	×	×	×
SS, VSS, mg/l	3/week		×	×
Oxygen uptake, mg/l/day	3/week		×	
Dissolved oxygen, mg/l	daily		×	
Microscopic analysis (gram stain)	1/week		×	
Color, turbidity	3/week			
Significant ions, cmpds, etc.	3/week	×		×

[a] Sample to be withdrawn from influent feed line or raw waste containers.
[b] Sample to be withdrawn from the unbaffled tank.
[c] Sample to be withdrawn from effluent bottle.

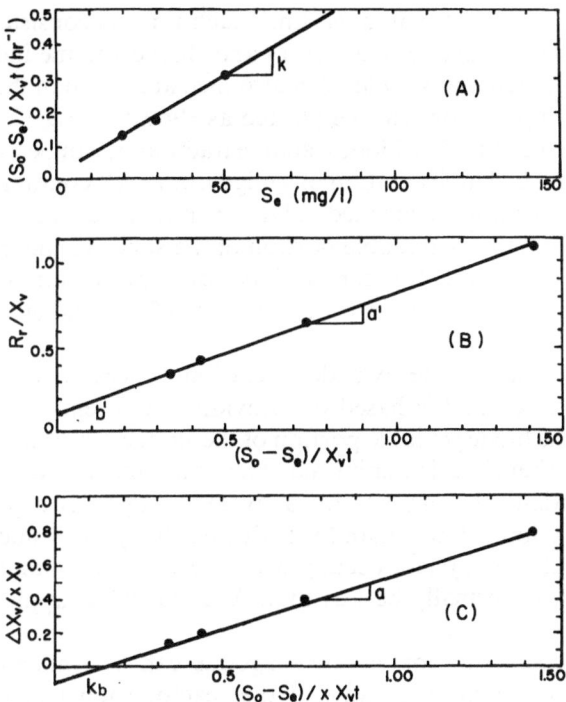

Fig. 12. Activated sludge design formulations

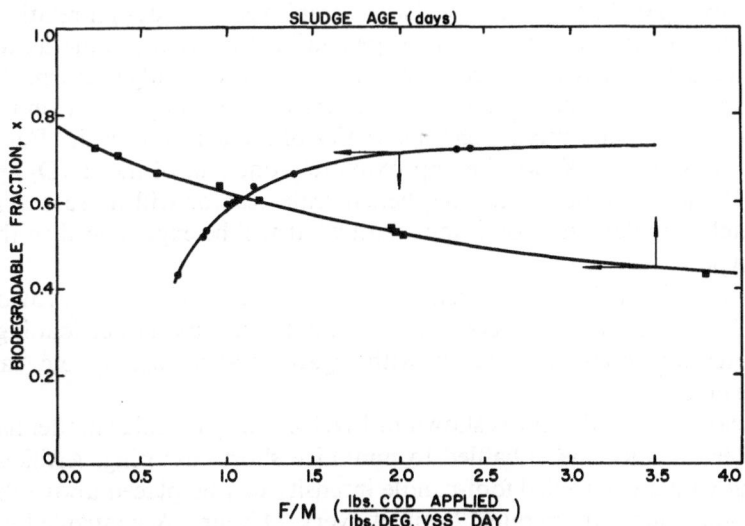

Fig. 13. Variation of degradable fraction with sludge age and with F/M for synthetic sewage

This schedule should be continued until such time as consistent results are obtained. The sludge settling curves and sludge volume index of the mixed liquor in each unit should be determined at the end of the test run. Data from each test series can be plotted as shown in Fig. 12 to obtain the desired coefficients. The biodegradable fraction, x, can be experimentally determined by batch aeration of sludge from each continuous treatment unit. The volatile suspended solids are measured as a function of time. The ratio of the degradable portion of the solids to the initial VSS present is the degradable fraction, x. A typical representation of biodegradable fraction as a function of loading rate (F/M) and sludge age is shown in Fig. 13 [11].

b) *Extended Aeration.* The extended aeration process is an activated sludge modification and is based on providing sufficient aeration time for oxidizing the biodegradable portion of the sludge synthesized. Sludge accumulation, therefore, is minimized. The obtainment of design coefficients is essentially the same as for activated sludge. As expected, the degradable fraction is lower than for activated sludge. The sludge age in an extended aeration system usually will exceed ten days, and the organic loading will normally be less than 0.20 lbs BOD_5 applied/day/lb. MLVSS.

c) *Aerated Lagoons.* Aerobic aerated lagoons can be simulated in the laboratory by removing the baffle of the reactor shown in Fig. 9 and operating at a flow-through condition. Pilot plants, geometrically similar to the anticipated prototype and operated at approximately the same power level (HP/1000 gals), may also be employed. The design relationships for substrate removal and oxygen utilization is the same as for activated sludge. Because there is no clarification or sludge return, the solids in the basin will approach an equilibrium level dependent on the organic loading and mixing characteristics of the aeration basin. For a soluble wastewater, X_v will be approximately one-half of the BOD_5 applied. It should be noted that the bench scale reactor will more nearly approach complete mix conditions than what will be experienced in the prototype.

d) *Waste Stabilization Ponds.* Bench scale ponds can be used to provide insight as to expected prototype performance at various surface loadings and detention times. Information with regard to algal toxicity, etc. can also be determined.

A typical bench scale unit is shown in Fig. 14. This particular model has a 45 liter capacity and is baffled to minimize short circuiting. Artificial lighting with a 600 to 800 foot-candle intensity can be placed above the ponds and timed to operate 12 out of every 24 hours. A controlled air stream is used to simulate wind and wave action. Distilled water is added to compensate for evaporation losses. Pond performance is evaluated by

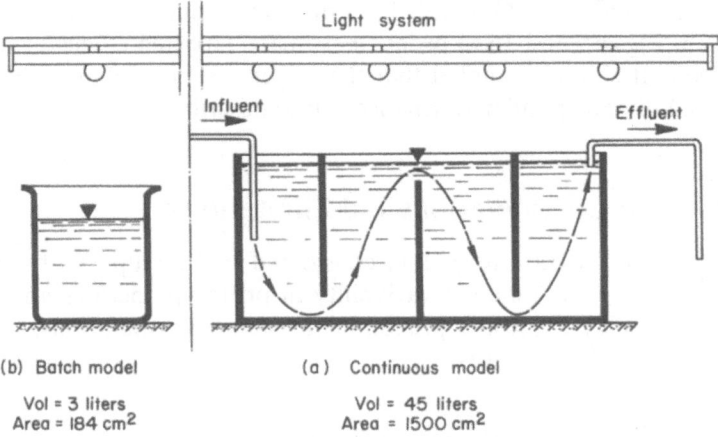

(b) Batch model   (a) Continuous model

Vol = 3 liters   Vol = 45 liters
Area = 184 cm²   Area = 1500 cm²

Fig. 14. Waste stabilization pond models

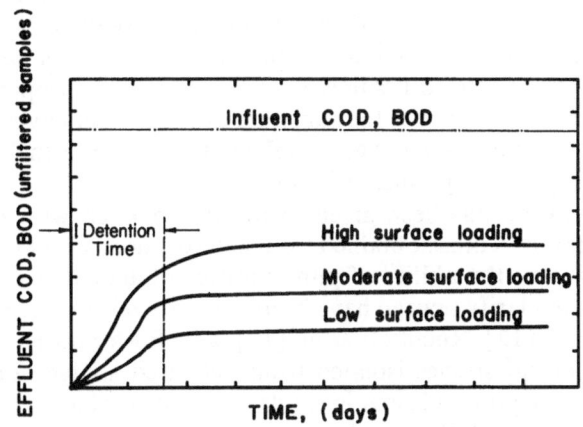

Fig. 15

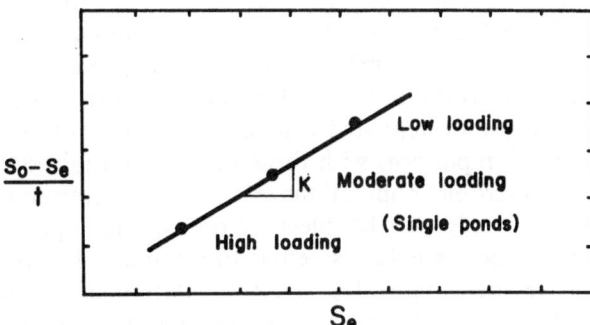

Fig. 16. Pond removal correlation

influent and effluent COD and BOD measurements. Data is plotted as shown in Figs. 15 and 16 in order to estimate expected efficiency of the prototype. It should be noted that the larger the pilot plant, the more accurate full-scale pond performance will be described.

2. Scale-Up Considerations and Limitations

Several factors which are seldom considered in scale-up due to added complexity and the inherent variability in prototype performance, but nevertheless play a role are: predominant bacterial populations, turbulence, and non-ideal reactors.

2.1. Influence of Bacterial Population

The value of the coefficients determined by laboratory experimentation will be dependent on the bacterial ecosystem of the reactor. Whether an identical biomass will predominate in the prototype is a matter of conjecture. It is reasonable to assume, however, that with proper seeding and acclimation, the nature of the substrate and other set environmental conditions (detention time, oxygen level, etc.) will dictate the predominant bacterial species, regardless of scale.

Some consideration has been given to results obtained with pure cultures as compared to those obtained from heterogeneous bacterial populations. Studies on the ability of pure cultures of activated sludge microorganisms to clarify sewage have been made by several investigators. Butterfield et al. [12], Ruchhoft et al. [13], and Buck and Keefer [14], employing zoogleal species isolated from activated sludge, found that BOD removal by pure cultures is similar to that attained from mixed flora of whole sludge. Other workers reported that no pure culture was able to clarify domestic wastes as efficiently as sewage-seeded sludge. Some species though did approach 90% efficiency. Van Gils, comparing the oxidation of selected organic compounds by pure cultures and by whole sludge, observed that members of the genera "Pseudomonas, Arthrobacter, Flavobacterium, and Zooglea" gave oxygen consumption curves similar to those of activated sludge [15]. No work comparing whole sludge oxidation patterns with those of pure and mixed cultures, using the ratios of microbial populations existing in an aeration tank has been reported. Knowledge of the effect of one species upon another species or group of species is fundamental to attaining the goal of the systems analyst, i.e., development of a representation of the activated sludge aeration tank environment as a predictable functioning ecosystem. Further research is obviously required in this area.

2.2. Effect of Turbulence

The agitation employed in bench studies, because of the influence of scale, is more severe than that encountered in the prototype. This point becomes significant when one considers the bacterial flocs or clumps purifying the wastewater. Numerous investigators [16, 17, 18] have shown that under the mixing and aeration conditions in the activated sludge process as conventionally employed, the biological flocs (50 to 200 µ) may be composed of aerobic surface layers with an anaerobic center. The aerobic fraction is much more efficient in substrate removal. A system, therefore, with low turbulence and consequently large floc diameters will have a reaction rate which is diffusion limited. Exposure of more biomass surface area by increased turbulence should increase floc activity by reducing the resistance to oxygen and nutrient penetration. There is an optimum turbulence level set by the settleability of the flocs and the economics of the system.

Rickard and Gaudy [19] showed the relationship between power level, organic removal, oxygen uptake rate, and the endogenous rate, b. Busch [20] has recomputed their data and showed that increasing the hydraulic shear intensity from 300 to 1000 resulted in an increase of organic removal rate from 0.3 gm COD/gm MLSS/hr to 0.7 gm COD/gm MLSS/hr and an increase in the endogenous coefficient, b, of 0.14 to 0.20. It is not possible, then, to relate experimental results to common field operating conditions, since the hydraulic shear rate does not readily relate to the power level in aeration basins expressed as HP/1000 gal. Degree of mixing is very difficult to measure in completely mixed basins and to scale-up to prototype units. The mixing regime is influenced by variation of temperature, air-flow rate, the type of air injector or agitator, agitator speed, mean residence time, tank geometry, solids level, and entrance and exit conditions. Mixing intensity may be evaluated by calculating dimensionless units (i.e. Reynolds Number), power relationships (HP/1000 gals), or oxygen absorption rates ($K_L a$).

Bartholomew [21] considering scale-up of submerged fermentations concluded that the most usable concept for scale-up is on the basis of oxygen transfer. The overall mass transfer coefficient, $K_L a$, will reflect air and flow agitation and thus can relate the reactor physical performance to productivity. Employing $K_L a$, then, has a more rational basis than HP/1000 gals. The oxygen absorption rate may be determined by the method outlined by Eckenfelder and Ford [5].

Many observers have correlated productivity for a variety of fermentations with oxygen transfer rate. Since $K_L a$ does vary with basin geometry, it is particularly interesting to note the work of Strohm et al. [22]. Fig. 17 correlates the yield of bakers' yeast to sulfide oxidation rates for

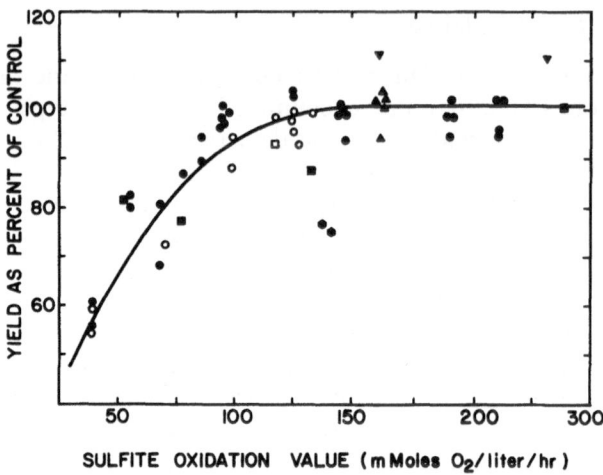

Fig. 17. Correlation of yield of yeast fermentation with sulfite oxidation value

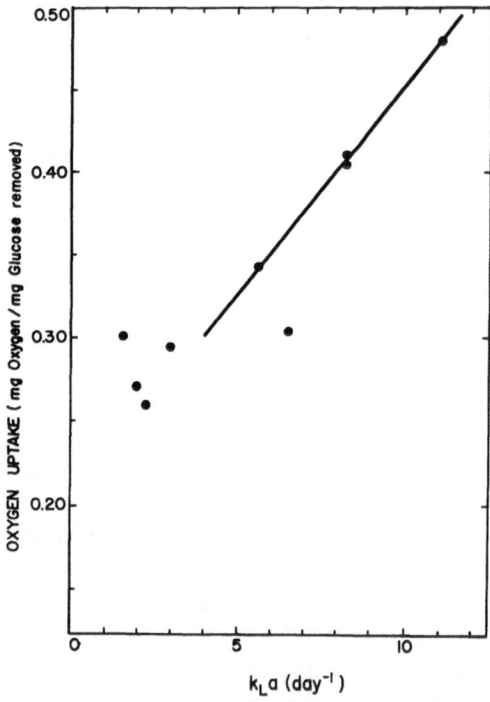

Fig. 18. Ratio of oxygen used to glucose removed as a function of turbulence as measured by $K_L a$

several sizes of geometrically dissimilar fermentors — both agitated and non-agitated vessels. The authors stated that this correlation was obtained over a tenfold range of values as long as the air rate for a specific yield was expressed in terms of linear flow or volume of air per volume of solution per unit of time. While, in fact, this was not a wastewater treatment system, the fundamental biological principles are identical and the same trend would be expected. Marlar [23] using batch reactors and a glucose substrate obtained the relationship shown in Fig. 18 between cumulative oxygen uptake and K_La. This plot supports the concept of increased activity at increased turbulence intensities.

In summary, turbulence is an important criterion in scale-up. While it is not easy to measure or apply to scale-up, it's effect might be approximated by conducting laboratory studies at various mixing intensities as measured by the oxygen absorption rate. Coefficients obtained at the K_La value anticipated for prototype performance should be employed in design.

2.3. Non-Ideal Reactors

The design equations previously developed are based on completely mixed conditions. While most activated sludge and extended aeration plants will reproduce the classic completely mixed tracer washout pattern, aerated lagoons and waste stabilization ponds exhibit neither plug flow nor a completely mixed system. The resultant intermediate system may be described by conducting tracer tests according to the method proposed by Murphy and Timpany [24] and Thirumurthi [25]. It should be noted that the complete mix approach is inherently the most conservative.

3. Conclusions

Laboratory or pilot plant data can be used to determine:
1. General response characteristics to be employed in the evaluation of alternate treatment schemes.
2. Parameters and coefficients for process design. These relate to the kinetic rates of reaction, quantity of sludge produced, and the amount of oxygen and consequently horsepower required.
3. Operational problems to be expected. These include nutrient requirements, toxic effects, sludge settling and separation problems, temperature effects, and foaming problems.

Flexibility should be paramount in design to account for model limitations and uncertainties.

D. Application of Model to Available Data

Data correlating bench scale results with those of the constructed prototype for wastewater treatment facilities were reported by Mancini and
Barnhart [26] and Quirk [27].
Mancini and Barnhart in Figs. 19, 29, and 21 present the laboratory
results, the prediction line, and the operating results from a prototype
aerated lagoon system treating an industrial waste. The kinetic and
sludge production plots show reasonably good agreement between predicted and observed results.
Quirk, using the same mathematical model as Mancini and Barnhart,
compared BOD removals for pilot plant (two 55-gallon aeration basins)

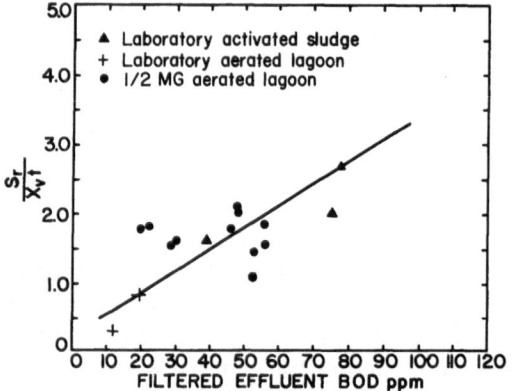

Fig. 19. Filtered effluent BOD vs. BOD removal rate

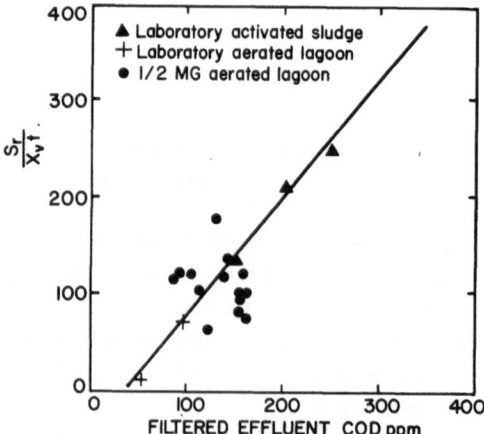

Fig. 20. Filtered effluent COD vs. COD removal rate

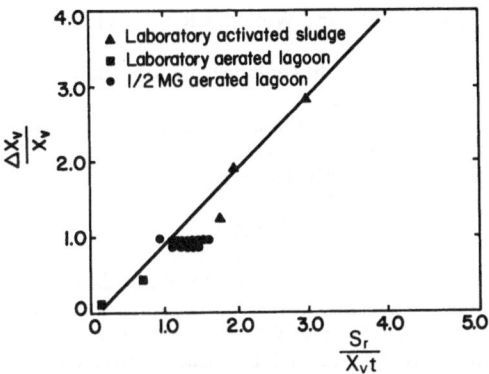

Fig. 21. Suspended solids production as a function of BOD removal

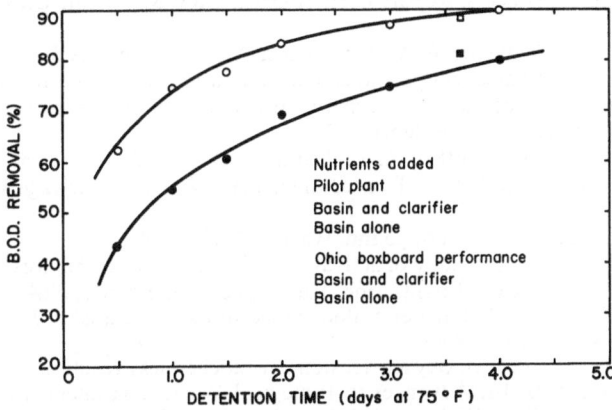

Fig. 22. Effects of detention time, nutrient use, and final clarification on BOD removal in pilot plant and board mill full-scale operation.

and full-scale operation of a boxboard wastewater as shown in Fig. 22. Full-scale basin performance depended upon the average temperature of the aeration basin, the use of a clarifier following the aeration basin and the addition of nutrients.

Since the degradable fraction refinement is currently being published, neither investigator employed it in his analysis. It should be noted, however, that even with its omission, excellent correlation for both cases was obtained.

References

1. McKinney, R. E.: Trans. ASCE **128**, Part III, 497 (1963).
2. McKinney, R. E., Ooten, R. J.: Concepts of complete mixing activated sludge. Transactions, 19th Annual Conference on Sanitary Engineering, Bulletin of Engineering and Architecture No. 60, University of Kansas, Lawrence (1969).

3. McKinney, R. E., Benjes, H. H., Jr., Weight, J. R.: J. WPCF **42**, 5 Part I, 737 (1970).
4. Eckenfelder, W. E., Jr.: Water quality engineering for practicing engineers. New York: Barnes & Noble, Inc. 1970.
5. Eckenfelder, W. E., Jr., Ford, D. L.: Water pollution control experimental procedures for process design. Austin, Texas: Jenkins Book Publishing Co. 1970.
6. Goodman, B. L.: Manual for activated sludge sewage treatment. Westport, Conn.: Technomic Publ. Co. 1971.
7. Goodman, B. L.: Design handbook of wastewater systems: Domestic, industrial, commercial. Westport, Conn.: Technomic Publ. Co. 1971.
8. Ford, D. L.: Evaluation of laboratory and pilot plant data for process design. Wastewater Treatment Process Design for the Chemical Industry, Proceedings of Seminar. Nashville, Tenn.: Vanderbilt University 1970.
9. Umbreit, W. W., Burris, R. H., Staffer, J. F.: Manometric techniques. Minneapolis: Burgess Publ. Co. 1959.
10. Eckenfelder, W. W., Ford, D. L.: Laboratory and design procedures for wastewater treatment process. Report EHE-10-6802, CRWR. Austin: University of Texas 1968.
11. Barnard, J. L., Eckenfelder, E. W., Upadhyaya, A. K., Englande, A. J.: Design optimization for activated sludge and extended aeration plants. To be published in the Proceedings of the 6th International Conference of Advances in Water Pollution Research — Jerusalem (1972).
12. Butterfield, C. T., Ruchhoft, C. C., McNamee, P. D.: U.S.P.H.S. **52**, 387 (1937).
13. Ruchhoft, C. C., Butterfield, C. T., McNamee, P. D., Wattie, E.: Sewage Works J. **11**, 195 (1939).
14. Buck, T. C., Keefer, C. E.: Sewage Ind. Wastes **31**, 1267 (1959).
15. Van Gils, H. W.: Bacteriology of activated sludge. Report No. 32, Research Institute for Public Health Engineering, The Hague, Netherlands (1964).
16. Mueller, J. A.: Normal diameter of floc related to oxygen transfer. J. San. Eng. Div., SA 2, 4756, April (1966).
17. Wuhrmann, K.: Effect of oxygen tension on biochemical reactions in sewage purification plants. In: Advances in Biological Waste Treatment. In: Eckenfelder, W. W., McCabe, W. J. (Eds.) Oxford: Pergamon Press 1963.
18. Pasveer, A.: Sewage Ind. Wastes **26**, 28 (1954).
19. Rickard, M. D., Gaudy, A. F.: WPCF **40**, R129 (1968).
20. Busch, A. W.: Aerobic biological treatment. Houston, Texas: Oligodynamics Publishing Company 1971.
21. Bartholomew, W. H.: Appl. Microbiol. **2**, (1960).
22. Strohm, J., Dale, H. F., Peppler, H. J.: Appl. Microbiol. **7**, 235 (1959).
23. Marlar, J. T.: The effect of turbulence on bacterial substrate utilization. Report WRC-0568, Georgia Institute of Technology, Atlanta, Georgia (1968).
24. Murphy, K. L., Trimpany, P. L.: J. San. Eng. Div., SA5, 5496, October (1967).
25. Thirumurthy, D.: Design principles of waste stabilization fonds. J. San. Eng. Div., SA2, 6515, April (1969).
26. Mancini, J. L., Barnhart, E. L.: Advan. Water Quality Improv. **1**, (1968).
27. Quirk, T. P.: Water Wastes Engr. **6**, D-1, July (1969).

Professor W. Wesley Eckenfelder, Jr.
A. J. Englande
Dept. of Environmental and Water
Resources Engineering
Vanderbilt University
Nashville, TN 37303/USA

Brian L., Goodman
Black and Veatch,
Consulting Engineers
Kansas City, MO/USA

CHAPTER 5

Cellulose as a Novel Energy Source

E. T. REESE, MARY MANDELS, and ALVIN. H. WEISS

With 8 Figures

Contents

Cellulose was undoubtedly man's first fuel, being the major component of wood for his fires. Therefore, when we consider cellulose as a "novel" energy source we have more indirect use in mind, such as its conversion to synthetic fuels (Huge, 1968) and to food (Meller, 1969).

1. Availability of Cellulose

Approximately 0.1% of the solar energy incident on earth is fixed by green plants through photosynthesis. The annual net yield (excess of photosynthesis over respiration) of this process amounts to about 15—20×10^{10} tons of organic plant substance. Half of this is cellulose (Tracey, 1964; Bellamy, 1969). Since one man requires 500 g of food (70 g protein, 80 g fat, 350 g carbohydrate) per day to maintain himself (Klicka, 1970) at his energy output of 130 watts (Tracey, 1964), the present world population of 3.5×10^9 requires more than 5×10^8 tons of food per year and the demand is steadily increasing. Since much of the plant is inedible, unavailable, or eaten by animals, food shortages are developing, and will soon become acute. We can look to the vast, annually renewed, cellulose as a source of food for man, as a substrate for single cell protein, and as a raw material for fermentation. It is astonishing that cellulose has been practically unused for these purposes, although it is much used as paper, textiles, wood, and as a source of chemicals. The only significant use of cellulose as food is as part of the fodder for ruminant animals (Baumgardt, 1969; Pigden and Heany, 1969; Virtanen, 1966).

But, plentiful as cellulose is as a primary product of nature, it is the waste cellulose, the unwanted by-product or the no longer useful end item, that offers the most immediate promise for economic utilization. Cellulose is a major component of (a) agricultural wastes (straw, stubble, leaves and stalks of many plants, rice and other hulls, peanut, almond and other shells, corn cobs, bagasse, etc.), (b) food processing wastes (fruit peels, pulp, coffee grounds, pomace, vegetable trimmings, etc.), (c) wood wastes (brush, chips, bark, sawdust, paper mill fines, etc.), and (d) municipal wastes (40—60% of solid wastes chiefly as garbage and waste paper). The problems of disposing of these wastes by composting, use as manure, land fill, or burning, without undue pollution are acute, expensive, and greatly increase the incentive to use cellulose as an energy source (Cellulose solids wastes seminar, 1969). The availability of cellulosic wastes ensures abundant cheap substrates for any processes that are developed.

2. Cellulose Chemistry

Cellulose is the most widely distributed skeletal polysaccharide, amounting to approximately fifty per cent of the cell-wall material of wood and plants. Some natural materials are practically pure cellulose, e.g. cotton. Cotton is α-cellulose, a form insoluble in 17.5% NaOH. Plant and wood celluloses generally contain β-cellulose as well, a material soluble in the indicated solution. Thus, although cellulose is formed from D-glucose building blocks joined by β-1, 4-glucosidic bonds, there are differences in the degree and types of association within cellulose molecules.

Wood cellulose occurs in the presence of hemicelluloses of related structure and with lignin, a nonpolysaccharide. Wood also contains oleoresins, which are commercially extracted and steam distilled to produce turpentine and rosin (naval stores). Wood chemistry, then, is significantly more complex than α-cellulose chemistry.

The cellulose molecule is a polymer with molecular weight generally in the range of 300000—500000. Table 1 shows typical materials and their degree of polymerization. The structural formula is shown below:

Table 1. Cellulose characteristics (Kirk and Othmer, 1964)

Source	Molecular weight	Degree of polymerization
Native cellulose	600000 – 1500000	3500 – 10000
Chemical cottons	80000 – 500000	500 – 3000
Wood pulps	80000 – 340000	500 – 2100
Rayon filament	57000 – 73000	350 – 450

If one considers conformational analysis, in which anhydroglucopyranose units are in the chair form, the structure is more correctly shown as follows:

Note the manner in which cellulose is built from glucose units. Glucose, as well as cellobiose, cellotriose, and cellotetraose can be isolated when cellulose is hydrolyzed. Complete hydrolysis by acid yields D-(+)-glucose as the only monosaccharide.

In general, other polysaccharides also occur in the presence of cellulose. For example, cereal straws and bran contain pentosans, (most commonly, xylan, which is built up from D-xylose units) which yield pentoses on hydrolysis, rather than glucose. Starch is present in the majority of plants, and this material is built up from maltose units.

The cellulose molecule is thread-like, existing as fibrils — long bundles of molecules stabilized laterally by hydrogen bonding between hydroxyl groups of adjacent molecules. Molecule arrangement in the fibrillar bundles is so regular (but not perfect) that cellulose has a crystalline X-ray diffraction pattern. Hydrogen bonding and the arrangement of the cellulose molecules in native cellulose (Liang and Marchessault, 1959) is shown on Fig. 1. An excellent discussion and bibliography of review articles on this subject is available (Ward, 1969). He points out that the consequence of the high degree of order in native cellulose is that not even water molecules, let alone enzymes, can enter the structure. Consequently, native cellulose is inert in the digestive tract. The structure of acid or alkali swollen cellulose is open; and this material is readily split by cellulase. Table 2 lists compositions of wood hydrolyzates (Underkofler and Hickey, 1954).

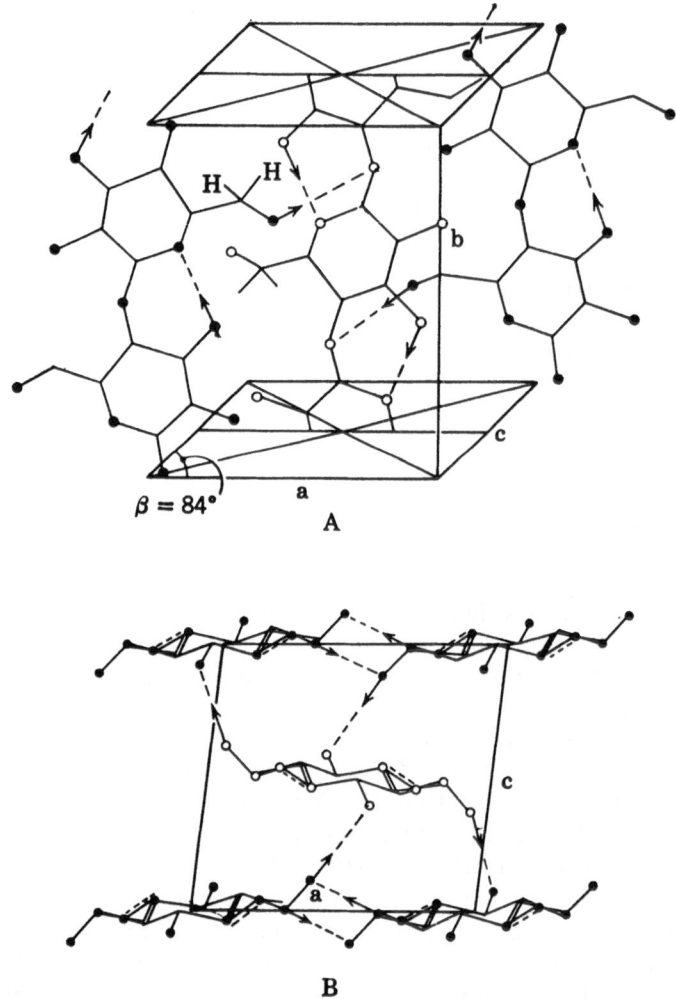

Fig. 1. Hydrogen bonding and the arrangement of cellulose molecules in native cellulose (Liang and Marchessault, 1959)

Table 2. Composition of the total hydrolyzate of wood (Underkofler and Hickey, 1954)

	Birch (in %)	Jack pine (in %)
Glucose	67.7	67.6
Mannose	1.8	14.1
Galactose	0.0	6.2
Fructose	–	–
Xylose	30.1	8.9
Arabinose	0.4	3.2
Total	100.0	100.0

3. Cellulose as a Source of Fuel

We are all familiar with the use cellulose has as an energy source. Wood has been a fuel since time immemorial — e.g. waste wood in sawmill operations is 50% of that processed. Unfortunately, the moisture content of wood is generally high — typically 50% — and this detracts from its heating value, since heating value decreases linearly with moisture content.

The heating value of cellulosic fuels does not approach that of fossil fuels. For example, the heating value of dried wood or bagasse is typically 8000—9000 Btu/lb. This can be compared to heating values of fuel oil, coke, and bituminous coal — approximately 19000, 13000, and 15000 Btu/lb, respectively (Fryling, 1966).

In addition to bagasse, other cellulosic materials besides wood are burned directly for their heating value. Municipal and industrial solid waste, black liquor, coffee grounds, rice hulls, and furfural residue, are examples of fuels that have been used for steam generation from waste heat boilers (Fryling, 1966).

Pyrolysis technology for upgrading cellulose to a more efficient fuel is ancient. Carbonization of hard wood to charcoal was probably the world's first chemical process; and it enabled cave men to have a smokeless fuel for inside utilization. Typical hardwood yields as wt. % obtained by retorting are charcoal 25.2, methanol 1.9, acetic acid 2.9, tar oil 5.0, gas 18.3, and water 46.7. Literature (Shreve, 1967) provides these figures as well as process descriptions of both hardwood distillation and the naval stores industry, which is based on soft woods. Charcoal is an unimportant energy source at present, having limited use for iron smelting and as a cooking fuel.

However, cellulose pyrolysis technology is being revitalized with the advent of municipal solid waste disposal problems. The following composition of an average municipal refuse is reported (Bell, 1964).

Component	Weight %
Moisture	20.73
Cellulose, sugar, starch	46.63
Lipids (fats, oils, waxes)	4.50
Protein	2.06
Other organic (plastics)	1.15
Ashes, metal, glass, etc.	24.93

The heating value of the overall refuse is 4917 Btu/lb, that of the moisture and ash-free portion 9048 Btu/lb. Organic refuse is not greatly different from wood in its character, and Fig. 2 shows that refuse growth has developed cellulose into a major and, unfortunately, readily accessible raw material in the United States.

Fig. 2 is taken from a published work (Rosen *et al.*, 1970). They project a yield structure of 22.5% char, 19.5% organic liquids, and 13.5% gas with the pyrolysis processing scheme shown on Fig. 3. They calculate that the process is technologically and economically feasible for pursuit by a municipality, in spite of the fact that the liquid and gas produced are only marginally suitable as fuels. The liquid contains valuable organic chemicals for recovery and the 26% ash content char has a heating value of 10277 Btu/lb.

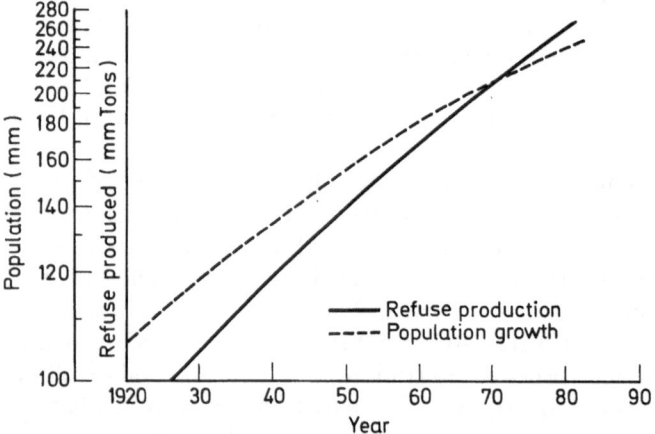

Fig. 2. Refuse production vs. U.S. population growth (Rosen *et al.*, 1970)

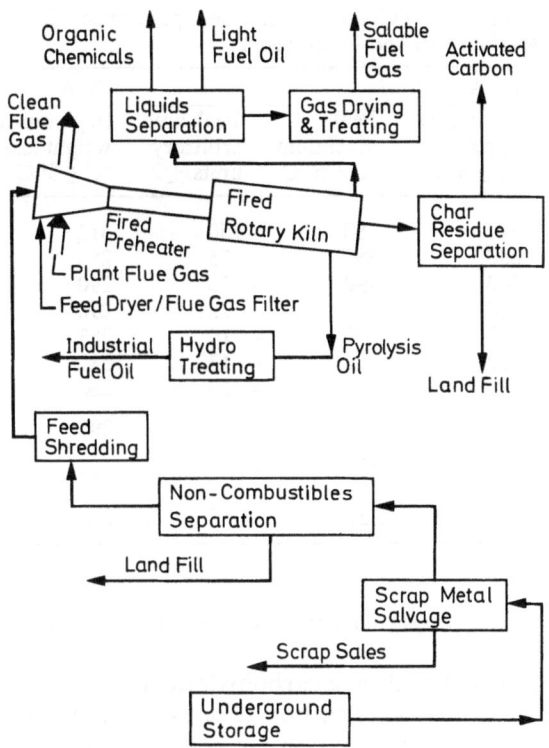

Fig. 3. Refuse pyrolysis plant (Rosen *et al.*, 1970)

Kinetic studies on the thermal degradation of both wood and cellulose in the presence of air or steam have indicated that the process can be treated as simple first order (Stamm, 1956). Other workers (Halpern and Patai, 1969) in a series of papers which provide a useful survey of cellulose decomposition, have confirmed by differential thermal analysis-thermogravimetric analysis (DTA-TGA) techniques Broido's observations (Kilzer and Broido, 1965) that two types of reaction occur. The first is endothermic, near 360° C, and results mainly in tar formation — loss of levoglucosan from the cellulose. The second is exothermic and corresponds to formation of water, carbon oxides and char. The effect of various additives on the decomposition rate of α-cellulose *in vacuo* is shown in Table 3. The amount of residue increases with additive basicity — endothermic reactions to produce tar predominate with no additive or with acidic additives.

The economics of producing liquid fuels from solid waste cellulose are far more attractive than those for producing a solid char, and the United States Bureau of

Table 3. Results of DTA-TGA-determinations performed simultaneously on α-cellulose samples (Halpern and Patai, 1969)

Additive	DTA			TGA	
10^{-4} mole/g	Beginning of peak °C ± 3° C	Peak maximum °C ± 5° C	Peak area arbitrary units ± 2	Beginning of weight loss °C ± 5° C	Residue at 400° C % ± 2%
None	310	380	− 150	312	13
NaHSO$_4$	280	350	− 75	280	16
Na$_2$SO$_4$	260	390	− 76	290	21
0.1 × Na$_2$CO$_3$	330	387	− 23	290	16
0.2 × Na$_2$CO$_3$	330	382	− 19	283	20
NaH$_2$PO$_4$	320	358	− 13	287	31
NaOAc	343	370	− 10	260	23
NaCl I	345	370	− 7	290	23
II	375	385	+ 6		
Na$_2$CO$_3$ I	315	337	+ 5	240	27
II	355	375	+ 5		
Borax	265	350	+ 55	280	43

Mines is conducting major research in this area. We can develop the problem in the following manner:

Consider the process of heating a carbohydrate. It first passes through a caramelized state, ultimately to carbon. (For example, charcoal made by heating wood). The pyrolysis reaction can be written as

$$C_6H_{10}O_5 \rightarrow C_6 + 5H_2O$$

This is an undesirable reaction from the standpoint of producing liquid hydrocarbons. Crude oil is higher-valued, the higher its hydrogen content[1], e.g. paraffinic vs. naphthenic crudes. In fact, the availability of hydrogen in refineries has led to processes such as hydrotreating, hydroforming and hydrocracking especially designed to add hydrogen to oil. As an example, an intrinsic part of the Shell process to convert 7° API Athabasca tar to 33° API sweet crude oil requires extensive hydrotreating facilities (Hoskins, 1964).

The only way for a liquid product to be formed from cellulose is for oxygen, rather than water, to be stripped from the molecule. This is evidently what occurred in Bureau of Mines work using both H$_2$ and carbon monoxide.

$$C_6H_{10}O_5 + 5 H_2 \rightarrow C_6H_{10} + 5 H_2O,$$
$$C_6H_{10}O_5 + 5 CO \rightarrow C_6H_{10} + 5 CO_2.$$

---

[1] In refinery jargon, the lighter the crude, the higher the API Gravity (the lower the specific gravity), the more useful the oil.

The stoichiometry of the above equations is absolutely speculative and undoubtedly an oversimplification of the facts. Concerning the exact chemical structure of C_6H_{10}, we shall not speculate at this time. Certainly, there is no question but that the C_6H_{10} can be upgraded by hydrogenation.

$$C_6H_{10} + 2\,H_2 \rightarrow C_6H_{14}.$$

Also, oxygenated compounds may undoubtedly form (e.g. phenols, cresols).

Table 4. Conversion of cellulose to oil (Appell *et al.* 1969)

Cellulose	Gas	Catalyst	Conversion (%)	Benzene-soluble product (wt. %)
Crude	CO		90	40
Ash-free	CO	None	63	17
Crude	H_2		60	20
Ash-free	CO	Na_2CO_3	94	46
Ash-free	H_2		90	27

Appell, Wender, and Miller have reported on their work at the Bureau of Mines (Appell *et al.*, 1970b; Appell *et al.*, 1969; Appell *et al.*, 1970a; Anon, 1969). Table 4 lists their results with cellulose charged to an autoclave containing water, pressured to 1500 psig with either CO or H_2, and then heated for two hours at 350° C. Yields as high as 46 wt. % benzene-soluble product were obtained.

Table 5. Conversion of municipal refuse to oil (Appell, 1970a) (20 min at 380° C, 1500 psig initial pressure)

Gas	CO	H_2
Conversion, %/wt.	94	80
Product distribution, %/wt.		
Oil[a]	41	18
Residue	6	20
Water	27 – 36	45
CO_2	15 – 24	15
Other	2	2

[a]Composition: C, 79.6 %; H, 9.5 %; N, 1.9 %; S, 0.13 %.

They also extended their work to municipal refuse, newsprint, and sewage sludge and developed a low-temperature version of the process operable at 250° C. Table 5 shows products obtained from municipal refuse. A summary of pyrolysis, hydrogenation, and incineration work currently proceeding at the Bureau of Mines on municipal refuse is also available for reference (Corey, 1970). The author estimates that a ton of dry refuse can yield approximately 2.4 barrels of oil — an exciting and economically promising figure.

4. Cellulose as a Source of Food

The biological utilization of cellulose depends upon its conversion to glucose, which, in turn, is often converted into other useful products. We may consider this process either as a) a series of conversions in which glucose is only a transitory phase, never accumulating and often scarcely detectable, and b) a single conversion in which glucose is accumulated in larger amounts, to be subsequently employed in all of the ways recognized for this most common and readily metabolized sugar.

a) Multiple Conversions: cellulose → food (protein)

(aa) Cellulose → meat. The conversion of cellulose to meat by ruminants proceeds through a series of steps made possible by the microbial fermentation tank known as the rumen. Here cellulose → glucose → fatty acids → amino acids → protein in a remarkable system which, unfortunately, is limited to a very few animal species, not including man. For the most part, the cellulosic foods are grasses and other succulent plants. Efforts are continually being made to modify cellulosic wastes so that they can be substituted for the normal foodstuffs. Most interesting information is provided by one of the prime promoters of this process (Virtanen, 1966). This is an agricultural problem outside the scope of the present discussion.

(bb) Cellulose → microbial protein. Many microorganisms grow well on cellulose as a sole carbon source. These include filamentous fungi and aerobic bacteria from soil, anaerobic bacteria from rumen and sewage, actinomycetes from compost, and wood-rotting basidiomycetes. It would seem that some of these should be suitable sources of food, but, except for a few varieties of mushrooms used more for flavor than nutrient, no such microorganisms are eaten, and no commercial fermentations based on cellulose are used to produce food products. Microorganisms are, of course, eaten in many fermented foods, but these organisms grow on sugar, starch, or protein, and not on cellulose and do not increase the amount of available foodstuff.

However, microorganisms can serve as food for animals, and presumably for men. Most of the research has been in the direction of converting

Table 6. Production of protein by *Candida utilis* grown on acid hydrolyzates of wood (Harris, 1949)

Hydrolyzate of	Rate of feed per hr (Litres)	pH	Yeast yield %	Protein content %	Sugar used %
Western larch	3	5.3	45.9	49.0	91.2
Western larch	3	6.3	47.4	47.4	91.7
Douglas fir	3	5.0	42.6	45.3	95.1
Lodgepole pine	3	5.9	51.7	53.8	94.0
Southern yellow pine	3	5.5	45.0	52.1	92.1
Aspen	3	5.5	45.2	52.7	94.8
Southern red oak	3	5.5	49.3	54.0	83.8

readily available carbohydrate (but not cellulose) into microbial fat or protein (Table 6) by yeasts, and in showing that such yeast can be incorporated profitably into animal diets (Bressani, 1968). Other fungi and bacteria have received much less attention. Gray has been a strong advocate of fungi as a source of protein using sugar or starch in substrates such as cane molasses, sugar beets, potatoes, sweet potatoes, manioc, corn and rice. Fungi are easy to grow, give 40—50% yields of cell matter (based on the substrate), and contain 13—38% protein (Gray, 1966). Some other workers (Litchfield et al., 1963) have grown several species of *Morchella* (a "mushroom") in submerged culture on glucose plus corn steep liquor, producing mycelium of 23—51% protein content with a nutritionally good amino-acid composition. Other fungi could be grown on cellulose but separation of the mycelium or fungal protein from undecomposed residues, and processing this into an acceptable food would present problems.

The investigation of the food value of specific cellulose-decomposing organisms has been extremely limited. Straw and hay were composted (Imre and Petch, 1967) for 21 days, pasteurized and inoculated with *Agaricus bisporus* (mushroom) which was allowed to grow for another 14—21 days. The resulting compost was proposed as a suitable fodder for various domestic animals. Cultures of rumen bacteria and protozoa were fed to rats and the protein was found to be nutritionally equal to 175% of its weight of soybean protein (Abdo et al., 1964). It was suggested that such cultures could be fed to monogastric animals. *Trichoderma viride* (Church and Nash, 1970) grown on waste from a corn-processing plant was fed to rats and its protein found equivalent to casein. A number of thermophilic cellulolytic bacteria are reported to have high rates of utilization of cellulose, and to produce protein of good amino-acid composition (Bellamy, 1969). The U.S. Department of

Health, Education, and Welfare supports a number of projects on utilization of cellulosic wastes through fermentation. Studies on submerged cultivation of fungi on waste paper and the protein value of the residues produced are also reported (Updegraff, 1969). Other investigators have reported (Srinivasan and Han, 1969) direct conversion of bagasse (sugar cane residue) to *Cellulomonas* cells. This organism grows well on bagasse which has been milled and extracted with alkali, and appears to produce a high quality protein. Development of a pilot plant to produce this bacterium in quantity sufficient for animal testing is also under way.

b) Single Conversions: cellulose → glucose

(aa) Acid Hydrolysis of Cellulose. The problems of cellulose utilization would be greatly simplified if cellulose could be converted into glucose, and the glucose then used as the substrate for further synthesis. Many attempts have been made to put this on an economically feasible basis. Wood wastes have been tested extensively since they are available in large quantities the year round. They have the disadvantages of containing lignins, phenols, pentosans and polysaccharides other than cellulose. Soft woods are preferable to hard woods because of their greater hexose content. The first commercial application was in 1913 at Georgetown, S.C., where a plant was built to hydrolyze southern pine mill waste by the American process in rotary steam-heated digesters with 2% sulfuric acid at 175° C. The dilute sugar solutions produced (25% yield) were fermented to 5000 gallons of ethyl alcohol per day. This plant and a second one at Fullerton, La., operated until 1923, when a decline in the price of blackstrap molasses made them unprofitable (Stamm, 1951). The German Scholler process used during World War II pretreated the wood with 1% HCl and then digested it with 0.5% sulfuric acid at 130°—190° C for 18—24 hours in stationary digesters for a sugar yield of 40—50%. The Bergius process, also used in Germany during World War II, used concentrated HCl in special acid-resistant equipment. It gave cleaner syrup, but was expensive because of the need to recover the HCl. The sugars from these processes were used to produce alcohol, and to grow *Candida* and *Oidium* yeasts for human food (Skoog, 1946; Srinivasan and Han, 1969). At the end of World War II, about 9000 tons of food yeast were being produced per year by this process.

The Forest Products Laboratory at Madison, Wisconsin, investigated the hydrolysis of wood to sugar during and after World War II. The Madison wood sugar process continuously percolated 0.4—0.6% sulfuric acid through chopped wood waste at temperatures gradually rising to 150—185° C and gave 4—5% sugar syrup in 45—55% yield from bark-free wood (40% yield if bark were included). The sugar syrup was

Table 7. Composition of wood sugar molasses (Harris, 1950)

Components	Weight %
Total solid matter	60 – 62
Reducing sugar (as glucose)[a]	48 – 50
Other carbohydrate	0.5 – 1.5
Non-sugar organic matter	6 – 8
Ash	2 – 3
Nitrogen	0.065
Volatile organic acids	1 – 2
Insoluble fiber	None

[a] The sugar in Douglas Fir molasses contains about 85% hexose and 15% pentose sugars. The sugar in a hardwood molasses, such as maple, contains about 65% hexose and 35% pentose sugars.

Table 8. World yeast production (Peppler, 1968)

	Bakers' yeast	Feed yeast
Europe	76 600 tons	125 500 tons
North America	61 000	37 500
The Orient	12 900	21 300
South America	6 300	1 200
Africa	2 350	2 200
World Total	159 150	187 700

neutralized with lime and concentrated to a 50% molasses (Table 7). Several hundred tons of wood-sugar molasses were made in a small pilot plant at Madison, and a larger pilot plant using the process was operated by the Tennessee Valley Authority. The molasses produced was fed to beef and dairy cattle with good results, used as a silage additive, and — less successfully — fed to sheep, swine, and poultry. It was also used to grow brewer's and baker's yeasts, and food yeasts such as *Candida utilis* (Table 6; Harris, 1950; Harris, 1949; Stamm, 1951).

None of the above processes has succeeded as a commercial operation since World War II. World demand for yeasts at present prices of 15— 29 cents per pound is low (Table 8) and easily met by yeasts produced on cane and beet molasses, sulfite liquor, whey, and fruit products (Peppler, 1968). Wood molasses for animal feeding has been unable to compete economically with cheap cane and beet molasses.

Nevertheless, the belief persists that sugars from cellulose will be needed in the future, and research on acid hydrolysis of cellulose continues,

Table 9. Effect of physical treatment on digestion of wood by
cellulase (Pew and Weyna, 1962)

Mechanical treatment	Residue from enzyme digestion %
Spruce	
Sliced 1 mm thick	99
Sawdust	94
Wiley Mill 80 mesh	93
Ground on vibratory mill 10 min	53
Ground on vibratory mill 8 h	28
Aspen	
Sliced 1 mm thick	99
Wiley Mill 40 − 60 mesh	94
Ground on vibratory mill 5 h	22

particularly in Iron Curtain countries. Much of the effort is devoted to
pretreatment of wood to make it readily and economically hydrolyzable
under mild conditions. A pertinent finding is that cellulose milled dry at
200—240° C is soluble in cold water and is readily hydrolyzed by dilute
acid (Krupnova and Sharkov, 1964).
(bb) Enzymatic Hydrolysis of Cellulose. Until recently, enzymatic hydrolysis of cellulose was too slow and too inefficient to compete with the
successful conversion of starch to glucose by a fungal enzyme (Underkofler, 1969). Current developments have changed the outlook. The two
important factors are (a) the importance of pre-treatment of the substrate, and (b) the development of much more active enzyme preparations. Fine grinding of cellulose greatly increased its susceptibility to
hydrolysis by enzymes (Table 9; Pew and Weyna, 1962; Ghose and Kostick, 1969). When cellulose pulp is heated and milled, a product is formed
from which high density suspensions (up to 50% w/v) can be prepared.
Without such a treatment, it is difficult to form a 10% suspension.
Processes in which solutions of high glucose content are produced elimi-
nate the cost of evaporating more dilute solutions. The enzyme system
which now has great promise is that from *Trichoderma viride* (Mandels
and Reese, 1964). This preparation can completely degrade even the
most resistant forms of cellulose, e.g., crystalline cellulose (Fig. 4). By
treating a heated and milled cellulose with a concentrated *T. viride* cellulase, Katz and Reese (1968) obtained a glucose syrup of 30% concentration (Fig. 5). Ghose (1969) and Ghose and Kostick (1969) studied this
process in a complete multi-stage continuous reaction system, using
heated and milled cellulose, and culture filtrates of *T. viride*. In a contin-

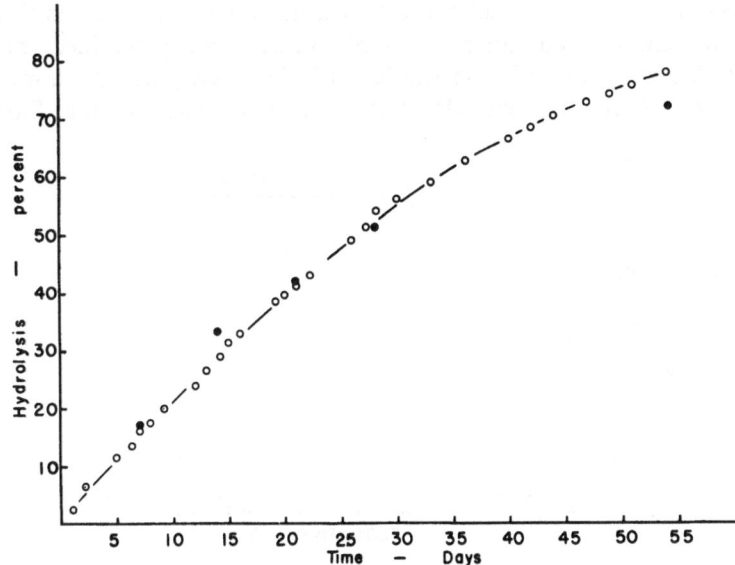

Fig. 4. Hydrolysis of cotton fiber by *Trichoderma viride* (Mandels and Reese, 1964) 29° C, pH 4.8, 1% cotton. ○ glucose production, ● weight loss

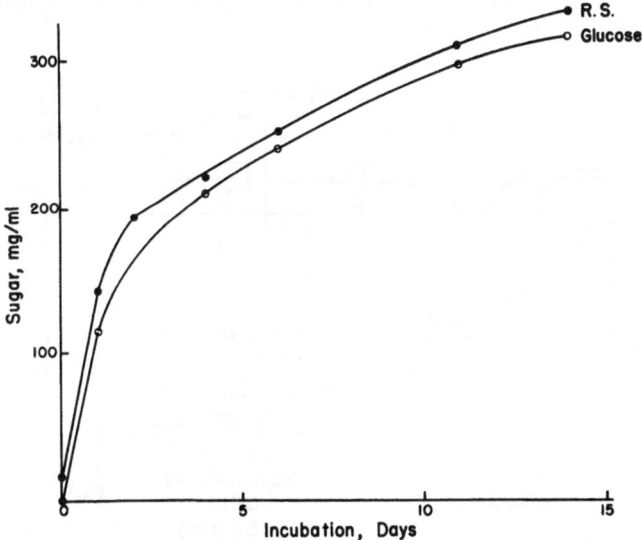

Fig. 5. Production of concentrated glucose solution by digestion of cellulose with enzymes (Katz and Reese, 1968). Substrate: Solka Floc 70 hours in Sweco mill, plus 30 min ball milling at 220 °C. Enzymes: Cellulase of *Trichoderma viride* 300 units/ml plus β-glucosidase of *Asperigillus luchuensis* 125 units/ml. Incubation: 50% suspension of substrate in enzyme, pH 4.5, 40° C. ●—● Reducing sugar as glucose; ○—○ glucose by glucose oxidase method

uous system with a 10% substrate concentration and a retention time of 40 hours, effluents containing over 5% glucose were obtained (Fig. 6). In subsequent studies (Ghose and Kostick, 1970) enzyme from a mutant strain of *T.viride* was used (Mandels *et al.*, 1970). These culture filtrates

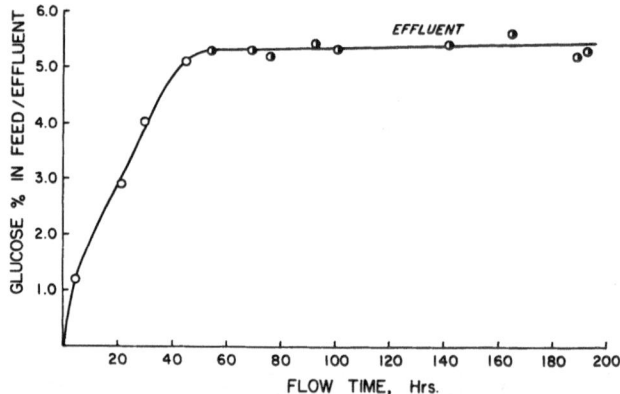

Fig. 6. Continuous saccharification of cellulose by enzymes (Ghose and Kostick, 1969). Substrate: Solka Floc, heated (200° C) and milled (76% < 53 μ). 10% suspension in enzyme pH 4.0—5.2. Enzyme: *T.viride* cellulase. Conditions: 4-liter, stirred tank reactor; dilution rate in continuous phase 0.025/h.; system became continuous at 50 hours.

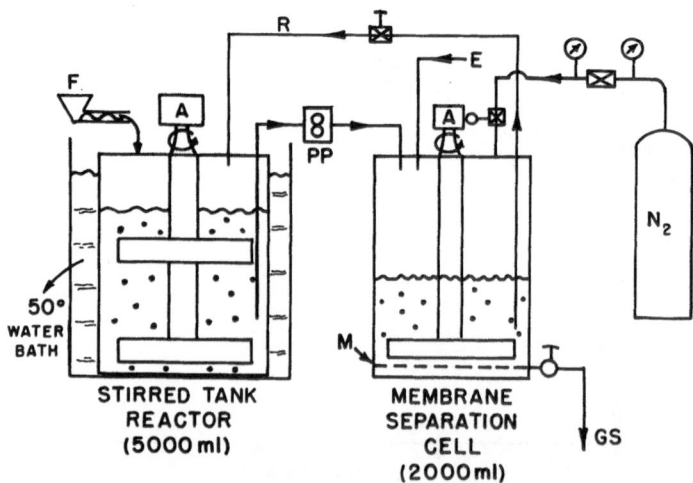

Fig. 7. Combined system for cellulose saccharification and removal of glucose in ultrafiltration cell (Ghose and Kostick, 1970). *A* Agitator. *PP* Peristaltic pump for transporting cellulose — cellulase slurry from reactor into the membrane cell. *GS* Glucose syrup free from cellulase and substrate. *F* Finely milled cellulose. *E* Solution of cellulase in water or in Tv medium. *R* Digest returning into reactor after glucose separator. *M* Mol. sieve membrane 10000—30000 mol. wt. cut off

Table 10. Material balance in the model system for cellulase saccharification and glucose removal (Ghose and Kostick, 1970)

Input			Glucose output		
Reaction time, h	Solid cellulose g	Water in cellulose ml	In digest initial %	In effluent average %	Removed g
0.00	900	3000	–	–	–
53.00	95	1135	14.21	9.59	108.8
75.00	125	2065	12.56	7.46	154.1
100.00	135	2020	13.40	6.74	136.2
172.25	100	2050	14.21	7.76	159.2
194.75	120	2000	13.71	7.26	145.1
220.25	100	1975	11.85	6.40	126.5
243.00	–	2025	12.10	6.30	127.5
Totals	1575				957.4

Milled cellulose ($< 25\,\mu$) was incubated with concentrated *T. viride* cellulase at pH 4.8, 50° C. At times noted 1/3 of the digest was pumped into an ultrafiltration cell and washed with water and dilute enzyme to remove glucose. After separation the slurry was returned to the reaction vessel and cellulose added equivalent to the glucose removed.

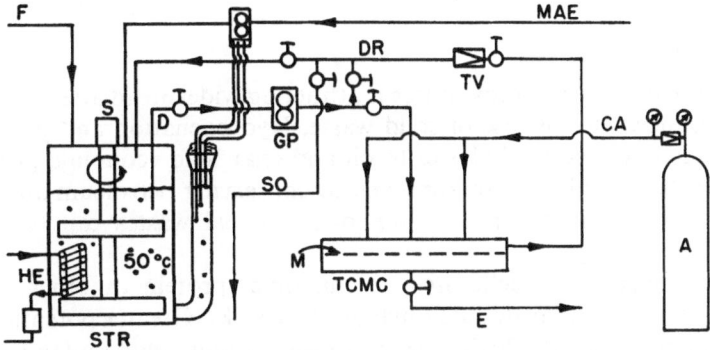

Fig. 8. Model for continuous enzymic saccharification of cellulose with simultaneous removal of glucose syrup (Ghose and Kostick, 1970). *S* Agitator. *GP* Gear pump transporting saccharified cellulose slurry. *D* Saccharified cellulose slurry transported into the thin channel membrane cell. *TCMC* Thin channel membrane cell separating sugars from the digest. *STR* Stirred tank reactor. *M* 20000—30000 cut off molecular sieve membrane. *HE* Heat exchanger to control reactor temperature. *E* Glucose syrup effluent free of enzyme or cellulose. *F* Finely ground cellulose feed. *DR* Saccharified slurry return from the membrane cell into the reactor. *MAE* Make-up enzyme (dilute) mixture of cellulase and Tv medium or water. *SO* 2—4% slurry output to cut down build-up of unreacted cellulose and denatured proteins. *TU* Throttle valve to control return of digest from the TCMC into the STR. *A* compressed air source. *CA* Compressed air line.

were concentrated 5—8 fold on polymeric molecular sieve membranes to further increase enzyme activity. Using concentrates in a 30% suspension of milled cellulose (<25 microns), in a batch process, it was possible to obtain solutions of 14% glucose in 50 hours. Application of the new ultrafiltration membranes made possible the development of a semicontinuous system, in which glucose passed through the sieves which retained enzyme and cellulose in the system (Fig. 7, Table 10). High rates of reaction were obtained, cellulose conversions approached 100% and the glucose was recovered as a clear aqueous syrup free of cellulose and protein. The freeze dried solids were over 80% reducing sugar. A continuous system was conceived (Fig. 8) in which (a) the enzyme and cellulose were retained in the reaction vessel, (b) the products were constantly being removed, and (c) water, make-up enzyme, and solid cellulose were constantly added to maintain the steady state condition (Ghose and Kostick, 1970).

These results suggest that efficient processes for enzymatic hydrolysis of cellulose at moderate temperature and pH may be achieved. Wilke and Rosenbluth (1969) of the University of California, Berkeley, have prepared preliminary cost estimates of the economics of using *T.viride* cellulase to produce glucose from wastepaper, and conclude that a 100 ton/day plant could be profitable.

5. Conclusions

Cellulose as a novel energy source is attracting widespread attention as a result of needs to dispose of solid waste. Hydrogenation and pyrolysis processes to convert cellulosics to oil and char hold economic promise and can be regarded as an approach to conserving petroleum and coal reserves. Even simple incineration of municipal wastes can result in useful heat energy.

Direct conversion of cellulose into microbial protein or other useful products has promise, despite obvious drawbacks. Greater potential lies in acid and enzymatic hydrolysis processes yielding glucose (and other sugars) from waste cellulose.

Research currently in progress will establish economically practical processes. The expertise of many disciplines will be needed to develop cellulose into a significant and novel source of energy in its various forms.

References

Abdo,K.M., King,K.W., Engel,R.W.: J. Animal Sci. **23**, 734 (1964).
Appell,H.R., Wender,I., Miller,R.D.: Conversion or urban refuse to oil. Bureau of Mines Solid Waste Program. Technical Progress Report 25 (1970b).

Appell, H. R., Wender, I., Miller, R. D.: Reprints Am. Chem. Soc. Fuel Division. No. 4, **13**, 35 (1969).

Appell, H. R., Wender, I., Miller, R. D.: Hydrogenation of municipal solid wastes with carbon monoxide and water. Presented at the National Industrial Solid Waste Management Congress, Houston, Texas (1970a).

Baumgardt, B. B.: Advan. Chem. Ser. **95**, 242 (1969).

Bell, J. M.: Proceedings national conference on solid waste research. American Public Works Assn., Special Report No. 29, p. 28 (1964).

Bellamy, W. D.: Cellulose as a source of single cell proteins. A preliminary evaluation. Report No. 69C-335. New York: General Electric Res. and Dev. Center, Schenectady 1969.

Bressani, R.: In: Mateles, R. I., Tannenbaum, S. R. (Ed.): The use of yeast in human foods, in: single cell protein, p. 90. Cambridge, Mass. M.I.T. Press 1968.

Cellulose solid wastes seminar. Dept. of Health, Education and Welfare, Cincinnati, Ohio (1969).

Church, B., Nash, H.: Use of fungi imperfecti in waste control. Report to USDA Federal Water Pollution Control Administration, Cincinnati, Ohio, 1970.

Corey, R. C.: Pyrolysis, hydrogenation, and incineration of municipal refuse. A Progress Report. Proceedings of the Second Mineral Waste Utilization Symposium, Chicago, Illinois, p. 299, 1970.

Fryling, G. R.: In: Combustion engineering. Chap. 14. Cambridge, Mass.: Riverside Press 1966.

Ghose, T. K.: Biotech. Bioeng. **11**, 239 (1969).

Ghose, T. K., Kostick, J.: Advan. Chem. Ser. **95**, 415 (1969).

Ghose, T. K., Kostick, J.: Biotech. Bioeng. **12**, 921 (1970).

Gray, W. D.: Advan. Chem. Ser. **57**, 261 (1966).

Halpern, Y., Patai, S.: Israel J. Chem. **7**, 673 (1969).

Harris, E. E.: Wood molasses for stock and poultry feed. Forest Products Lab. Report No. 1731. 22 p., 1950 revised 1960.

Harris, E. E.: Food yeast production from wood processing hydrolyzates. Forest Products Lab. Report No. 1754. 33 p., 1949 revised 1964.

Hoskins, D. A.: Hydrocarbon Process. Petrol. Refiner **43**, 122 (1964).

Huge, P.: Ann. Inst. Pasteur **115**, 574 (1968).

Imre, H., Petch, S.: In: Mushroom Science (Proc. Sci. Symp. Cultiv. Mushroom Inst. Congr. Mushroom Science 1965), p. 287 (1969).

Katz, M., Reese, E. T.: Appl. Microbiol. **16**, 419 (1968).

Kilzer, F. J., Broido, A.: Pyrodynamics **2**, 151 (1965).

Kirk, R. E., Othmer, D. F.: Encyclopedia of chemical technology. Vol. 4, 2nd Ed., 593. New York: John Wiley and Sons 1964.

Klicka, M.: Personal communication. U.S. Army Lab., Natick, Mass (1970).

Krupnova, A. V., Sharkov, V. I.: Gidrolizn i Lesokhim. Prom. USSR **17**, 3 (1964).

Liang, C. Y., Marchessault, R. H.: J. Polymer Sci. **37**, 385 (1959).

Litchfield, J. H., Vely, V. G., Overbeck, R. C.: J. Food Sci. **28**, 741 (1963).

Mandels, M., Reese, E. T.: Develop. Ind. Microbiol. **5**, 5 (1964).

Mandels, M., Weber, J., Parizek, R.: Appl. Microbiol. **21**, 152(1971).

Meller, F. H.: Conversion of organic solid wastes into yeast. Public Health Service Publication No. 1909 (1969).

Novel process could aid in waste disposal. Chem. Eng. News **43**, Nov. 17 (1969).

Peppler, H. I.: In: Single Cell Protein, p. 229, Mateles, R. I., Tannenbaum, S. R. (Eds.): Cambridge, Mass: M.I.T. Press 1968.

Pew, J. C., Weyna, P.: Tappi **45**, 247 (1962).

Pidgen, W. J., Heany, D. P.: Advan. Chem. Ser. **95**, 245 (1969).

Rosen, B. H., Evans, R. G., Carabelli, P., Zaborowski, R. B.: Economic evaluation of a commercial size refuse pyrolysis plant. Cities Service Oip Co., Cranbury, N. J., 1970.

Shreve, R. N.: In: Chemical Process Industries, 3rd Ed., Chap. 32. New York: McGraw-Hill 1967.

Skoog, F.: Food yeast production and utilization in Germany. Tech. Report, Office Quartermaster, 1946.

Srinivasan, V. P., Han, Y. W.: Advan. Chem. Ser. **95**, 447 (1969).

Stamm, A. J.: Proc. Am. Phil. Soc. **95**, 68 (1951).

Stamm, A. J.: Ind. Eng. Chem. **48**, 413 (1956).

Tracey, M. V.: Inst. Veterinary Inspectors. J. **28**, 31 (1964).

Underkofler, L. A.: Advan. Chem. Ser. **95**, 343 (1969).

Underkofler, L. A., Hickey, R. J.: Industrial Fermentations, Vol. 1. New York: Chemical Publ. Co. 1954.

Updegraff, D. M.: Degradation of waste Paper to Protein in Microbial Fermentations. Denver Res. Inst. Report on U.S.P.H.S. Research Grant (1969).

Virtanen, A. I.: Science **153**, 1603 (1966).

Ward, Kyle: In: Symposium on foods; carbohydrates and their roles, p. 55, Schultz, H. W., Cain, R. F., Wrolstad, R. W. (Eds.): Westport, Conn.: Avi Publ., Co. 1969.

E. T. Reese, Ph. D.
Mary Mandels, Ph. D.
Microbial Chemistry Group
Pioneering Res. Lab.
US Army Natick Laboratories
Natick, MA 01760/USA

Alvin H. Weiss, Ph. D.
Dept. of Chemical Engineering
Worcester Polytechnic Institute
Worcester, MA 01609/USA

CHAPTER 6

The Culture of Plant Cells

Mary Mandels

With 6 Figures

Contents

1. Introduction

Cells from higher plants have been cultured for about 30 years, a relatively short time compared to bacteria and other microorganisms. The primary objective was to gain a better understanding of plant growth and morphogenesis. Early efforts were towards finding tissues such as root tips, cambium, and tumors which were relatively easy to grow. The next phase established nutrient requirements especially for organic growth-regulating substances and led to successful culture of cells from all parts of the plant and representing most plant families (Carew and Staba, 1965; Gautheret, 1959; Hildebrandt, 1962; Murashige and Skoog, 1962; White, 1963). Cultures from monocotyledons were more difficult, but today cells of wheat (Gamborg and Eveleigh, 1968a; Shimada et al., 1969), barley (Gamborg and Eveleigh, 1968a), rice (Yatazawa et al., 1967), corn (Graebe and Novelli, 1966a and 1966b) and sugar cane (Nickell and Maretzki, 1969a) are being grown. The culmination of these studies was the regeneration of normal plants from callus cultures. Manipulation of nutrients and environmental conditions induced undifferentiated cells to produce specialized cells, organs such as roots or buds, and finally even embryos which could be grown to mature plants that flow-ered and set seed (Earle and Torrey, 1965b, Halperin, 1964a, 1964b, 1966; Steward *et al.*, 1964; Vasil *et al.*, 1965b).

These successes have attracted attention to other possible applications. Plant cell cultures have been used, or proposed for use, in the study of host-parasite relationships in mixed culture with plant pathogenic virus-es, bacteria, fungi, or even with nematodes or cropfeeding insects. They might also be used to screen substances for plant toxicity and for eval-uating potential herbicides or growth regulators. As a biochemical tool they offer a means of dissociating plants into their structural units and producing large quantities of cells of a single type.

A recent conference (Nickell and Torrey, 1969) focussed on uses of plant cell cultures in crop improvement. Regeneration of whole plants from callus cultures is now probably possible with any plant if enough effort is made. The use of mutagens (chemicals or irradiation) on such cultures offers potential for producing new varieties of plants for breeding pro-grams. Protoplasts of plant cells have been produced by removing the cell walls with cellulases of *Myrothecium verrucaria* and *Trichoderma viride*. Such protoplasts have been observed to fuse, and to develop new cell walls. This may allow the development of viable intergeneric hybrids (heterokaryons) between valuable food or fiber crop plants (Nickell and Torrey, 1969). Other uses discussed included culture of haploids from pollen, allowing fixation and analysis of genetic characteristics and achievement of homozygotes; and meristem cultures which permit rapid increase in virus-free clones, or of slow-growing plants that require vege-tative propagation (Morel, 1964).

Finally plant cell cultures might be a source of special plant products, and this is the reason for including this chapter in a book on biochemical engineering. Among the products identified in cultures are: alkaloids including nicotine, atropine, hyascine, tomatine, and candicine; amino-acids; proteins; enzymes including amylase, invertase, catalase, peroxi-dase, indoleacetic acid oxidase, polyphenol oxidase, protease, and pectin methylesterase; carbohydrates including starch; glycosides; organic acids; pigments including chlorophyll, carotenoids, xanthophyll; flavon-oids, and anthocyanins; phenolics; tannins; lignins; saponins; steroids; terpenoids; antibiotics against *Staphylococci* and *Mycobacteria*; and growth regulators (gibberellins) (Nickell and Maretzki, 1969). Other pos-sible products include flavors, odors, hormones, toxins, and even food. Since plant cells can be grown in suspension cultures that in many ways resemble cultures of true microorganisms, many of the techniques al-ready developed by the fermentation industry are directly applicable to growing plant cells on a large scale for use in industrial fermentations or in producing products of biochemical interest. To date no such fermentation has been used to prepare chemicals, even on a laboratory scale. This report attempts to indicate studies of interest to the bioengineer, to point out some of the special problems to be faced in culturing plant cells, and to cite articles in which useful methods and techniques are described.

2. Isolation, Establishment, and Maintenance of Cultures

The commonest type of culture, and the most likely to be used in fermentation is the callus culture which is a mass of undifferentiated cells. Callus tissue can be derived from almost any plant part such as root, stem, embryo, endosperm, cotyledon, pollen, or fruit, by placing a small (2—5 mm diameter) sterile piece of the desired plant part on an appropriate agar medium containing either coconut milk (Table 1) or suitable plant growth regulators (Table 2) to promote cell proliferation.

Sterile plant material can be obtained in a number of ways. The desired organ can be washed and surface sterilized with 3—5% sodium or calcium hypochlorite, 70% ethanol, or other antiseptic, rinsed in sterile water and bits of tissue removed aseptically from the interior. For firm tissues such as carrot roots a cylinder of tissue can be removed with a

Table 1. White's medium for growth of plant cell cultures (White, 1963)

Component	mg/l
$MgSO_4 \cdot 7 H_2O$	360
$Ca(NO_3)_2 \cdot H_2O$	200
$Na_2SO_4 \cdot 10 H_2O$	200
KCl	80
$NaH_2PO_4 \cdot H_2O$	16.5
$MnSO_4 \cdot 7 H_2O$	4.5
$ZnSO_4 \cdot 7 H_2O$	1.5
H_3BO_3	1.5
KI	0.75
Ferric tartrate	40
Glycine	3.0
Nicotinic acid	0.5
Thiamine	0.1
Pyridoxine	0.1
Sucrose	20000
pH 5.5	
Optional additives	
Agar 6 g/l	
Coconut milk 100 ml/l	

Table 2. Murashige and Skoog medium for growth
of plant cell cultures (Murashige and Skoog, 1962)

Component	mg/l
KNO_3	1900
NH_4NO_3	1650
$CaCl_2 \cdot 2\,H_2O$	440
$MgSO_4 \cdot 7\,H_2O$	370
KH_2PO_4	170
$MnSO_4 \cdot 4\,H_2O$	22.3
$ZnSO_4 \cdot 7\,H_2O$	8.6
H_3BO_3	6.3
KI	0.83
Molybdic acid	0.25
$CuSO_4 \cdot 5\,H_2O$	0.25
$CoCl_2 \cdot 6\,H_2O$	0.25
Sodium EDTA	37.3
$FeSO_4 \cdot 7\,H_2O$	27.8
Inositol	100
Glycine	2
Nicotine acid	0.5
Pyridoxine	0.5
Kinetin (6 furfurylaminopurine)	0.32
Thiamine	0.1
2.4-dichlorophenoxyacetic acid (2, 4D)[a]	0.5
Sucrose	30000
pH 5.5	
Optional additives	
Agar 6 g/l	
Phytone 1 g/l	

[a] naphthaleneacetic acid may be substituted 0.1 mg/l

sterile cork borer and sliced, discarding the end slices and using only the interior ones. Softer tissues can be cut or teased apart. Seeds may be sterilized for 15—20 min in 5% calcium hypochlorite, and planted (without rinsing) on agar medium containing nutrient salts. The media shown in Tables 1 and 2 are suitable if growth regulators are omitted. The seeds are germinated and grown for two weeks or longer in a greenhouse or lighted plant growth chamber. The sugar speeds up growth and aids in the detection of contamination. The sterile seedlings are then cut up and pieces transferred to agar medium for the initiation of callus tissue.

The plant pieces are incubated in the light or dark at 22—28° C. After 4—6 weeks callus tissue appears as a proliferation of cells which can be excised and transferred to fresh agar slants. Established cultures are

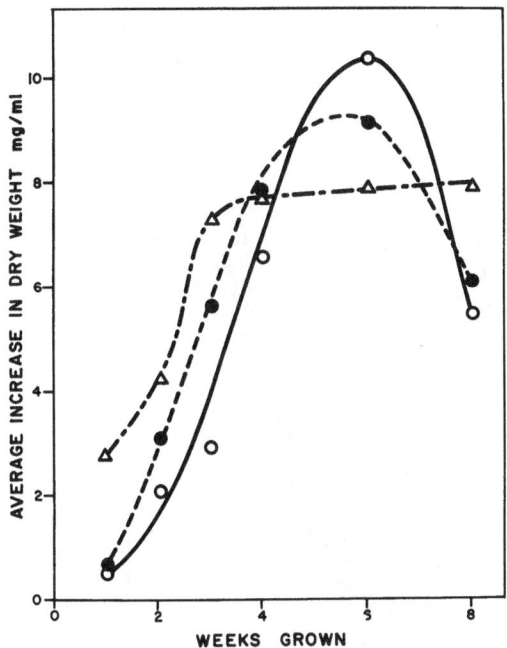

Fig. 1. Growth of callus cultures on agar (Mandels *et al.*, 1967). Cultures were grown in the dark at 28° C on Murashige medium (3% sucrose) and transferred at intervals of 1 – 8 weeks. O———O Carrot root, ●– – –● Lettuce leaf, △– - – -△ Bean leaf

maintained at 22—28° C and transferred at 3—6 week intervals. Growth of cultures is very slow (Fig. 1).

Suspension cultures in liquid media are initiated by transferring a piece of callus tissue from an agar slant into a shake flask and growing on a rotary or a reciprocal shaker. Initially these grow slowly, often as large tissue masses. If the callus is friable some cells may slough off and grow as free cells or as small masses. These can be separated by screening through gauze or wire mesh, or selectively transferred by means of a large opening pipette to establish cultures that grow well in suspension. Such cultures can be maintained over long periods of time by transferring (5—10% inoculum) every 2—4 weeks. Difficult tissues can be induced to grow in suspension by more elaborate techniques such as the use of special media, roller tubes, etc. Cultures so established may not grow well when transferred to the more rugged conditions of large scale fermentation. Cultures usually grow more rapidly in suspension than they do on agar (Fig. 2) but still very slowly in comparison to microorganisms such as bacteria.

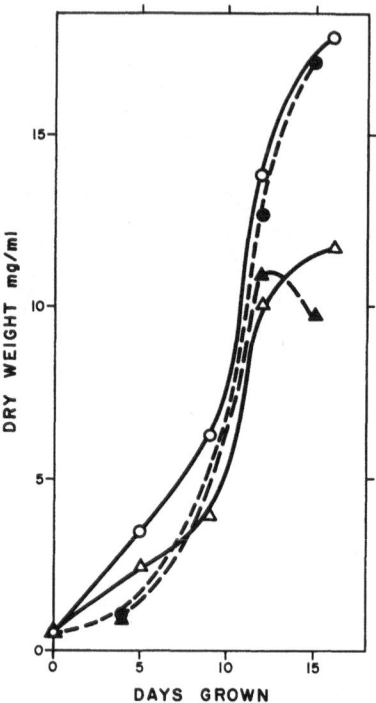

Fig. 2. Growth of plant cells in suspension cultures (Mandels *et al.*, 1967). Cultures were grown in the dark at 28° C on Murashige medium. O————O Bean leaf 3% sucrose, △————△ Bean leaf 2% glucose, ●– – –● Lettuce leaf 3% sucrose, ▲– – –▲ lettuce leaf 2% glucose

Techniques have been developed for the isolation of cultures derived from a single cell by (a) nurse cultures in which the single cell is nourished by an established tissue mass, but separated from it by a piece of filter paper (Muir *et al.*, 1958) or (b) growing isolated single cells in microdrops of conditioned medium (Jones *et al.*, 1960; Vasil and Hildebrandt, 1965a) or (c) plating out single cell suspensions (Earle and Torrey, 1965a; Gibbs and Dougall, 1963). The last method can give plating efficiencies of up to 100% so is particularly suitable for recovery of large numbers of single cells, as would be required in genetic studies or screening programs.

Many media have been developed for growing plant cell cultures, and many studies made of the optimum levels of various materials (Carew and Staba, 1965; Gamborg *et al.*, 1968; Gautheret, 1959; Hildebrandt, 1962; Mandels *et al.*, 1967; Mandels *et al.*, 1968b; Murashige and Skoog, 1962; White, 1963). In studies of morphological differentiation elaborate media containing coconut milk or other complex additives may be required. Many plant cells grow well on very simple media containing only inorganic salts, nitrate nitrogen, a carbon source such as sucrose, and trace levels of a few growth factors especially thiamine, an auxin such as 2,4-dichlorophenoxyacetic acid, and a cytokinin such as 6-furfurylaminopurine (Mandels *et al.*, 1968a; Gamborg *et al.*, 1968b). For special tissues or for the production of particular metabolites much effort may be required to develop a suitable medium.

Very few attempts have been made at long term preservation of plant cell cultures. One of the greatest problems facing workers in this field is the preservation and stabilization of their cultures. Plant cells have been frozen in the presence of 10% dimethyl sulfoxide with a 14% recovery (Quatrans, 1968) but most are maintained in active growth and transferred frequently.

3. Stability, Variation, and Differentiation of Cultures

Genetic instability of plant cell cultures is a serious problem (Nickell and Torrey, 1969a). Cultures frequently become polyploid, or they may permanently lose part or all of one or more chromosomes. Physiological variability that may or may not be genetic is frequently observed. Growth rates and cell type may show marked changes.

Requirements for accessory growth factors may also change. With long maintenance cultures tend to converge to a common cell type, to narrow their range of enzymes and metabolites, and to lose their ability to differentiate. Sometimes cultures become adapted to the medium on which they are maintained, and do not grow as well when transferred to a different medium. Even cultures originating from a single cell clone may show remarkable variation (Sievert and Hildebrandt, 1965). On the positive side variability and adaptation may sometimes be useful to the investigator.

For economic reasons and avoidance of contamination, strains of maximum growth rate are desired if the objective is production of cell mass or protein. Marked biochemical differences frequently exist between cultured cells and the tissues from which they originated (Tulecke *et al.*, 1962; Weinstein *et al.*, 1959; Weinstein *et al.*, 1962). The cultured cells are low in secondary metabolites such as pigments or alkaloids (Table 3). In the intact plant, these are produced by specialized cells that do not divide, whereas the fast-growing cultured cells are specialized for proliferation. Cell cultures can be induced to differentiate even in shake flasks (Mandels *et al.*, 1968b, Fig. 3). Differentiation is usually associated with a

Marigold

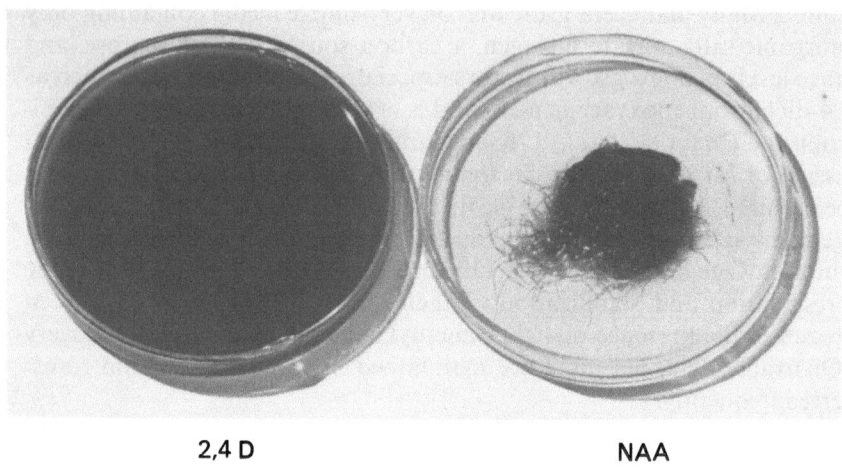

a

2,4 D   NAA

Carrot

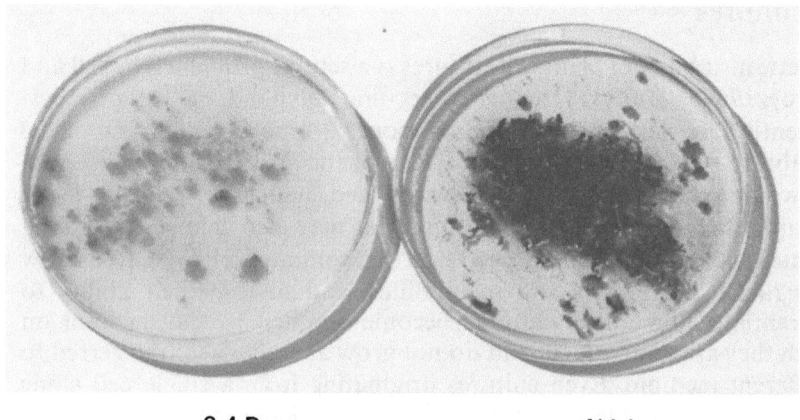

b

2,4 D   NAA

Fig. 3. Differentiation of plant cells in shake flasks (Mandels *et al.*, 1968b). Cultures maintained on Murashige medium with 2, 4 D produced roots (a Marigold) or plants (b carrot) when NAA (naphthalene-acetic acid) was substituted

decreased rate of growth but could result in increased production of some metabolites. The production of normal plants from cell cultures originating from various parts of the plant proves that the cultured cells contain all of the genetic information to produce any substance that the parent plant produces. Someday we will learn how to manipulate and release the regulatory controls that repress synthesis of desired metabolites without sacrificing the rapid growth of the cell culture. Until this has been achieved the production of fermentation chemicals may require a two-step process, the first for rapid increase in cell mass, the second for differentiation and production of secondary metabolites.

Another approach being investigated is to feed precursors of desired compounds to cell cultures. When ornithine and phenylalanine were fed at 0.1—0.2% to *Datura* cultures, alkaloid content of the cells increased, but growth was decreased so that total production of alkaloid by the culture was lowered (Chan and Staba, 1965, Table 3).

Table 3. Secondary metabolites in cell cultures compared to intact plants

Chlorophyll in *Nicotiana* (Laetsch and Stetler, 1965)	Chlorophyll mg/mg dry weight
Mature leaves	6.3×10^{-3}
Buds induced from callus	2.5×10^{-3}
Callus (pith)	2.7×10^{-4}
Alkaloids in *Datura* (Chan and Staba, 1965)	Alkaloids % dry weight
Intact plant	$0.206 - 0.306$
Callus (seed)	$0.015 - 0.056$
Callus (root)	$0.012 - 0.015$
Callus (stem)	$0.004 - 0.014$
Callus (leaf)	$0.006 - 0.021$
Suspension culture (leaf)	$0.009 - 0.054$
Suspension culture + 0.2% precursor[a]	$0.039 - 0.161$

[a] L-ornithine HCl or L-phenylalanine.

4. Techniques for Measuring Growth

Much of the original work on plant cell cultures was done by investigators who were more interested in type of growth than in growth rates. Results, even from studies on media optimization or establishing growth factor requirements were frequently presented as photographs of a series of callus cultures. Growth, especially on agar, is very slow, and frequently only a few cells in the culture divide so that growth may not be exponential. As a result few workers have expressed growth in terms such as specific growth constant, or generation time, that are familiar to the microbiologist.

For suspension cultures results are often presented on the basis of fresh weight as the simplest, easily reproducible measurement, and perhaps also because the numbers are larger. The relationship between fresh and dry weight changes as the culture grows (Fig. 4) and it is preferable to use dry weight. Some component of the cell such as protein or nucleic acid can be measured (Dougall, 1966). Cell counts (Wang and Staba, 1963, Fig. 5) are tedious to make and not very accurate because of the difficul-

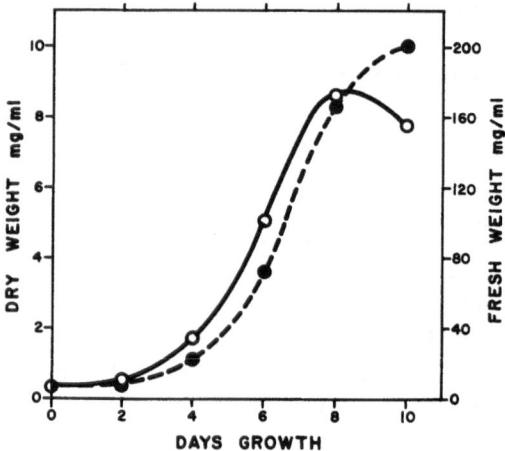

Fig. 4. Relationship between fresh and dry weights of cultured rose cells (Dougall, 1966). ●– – –● Fresh weight, ○———○ Dry weight

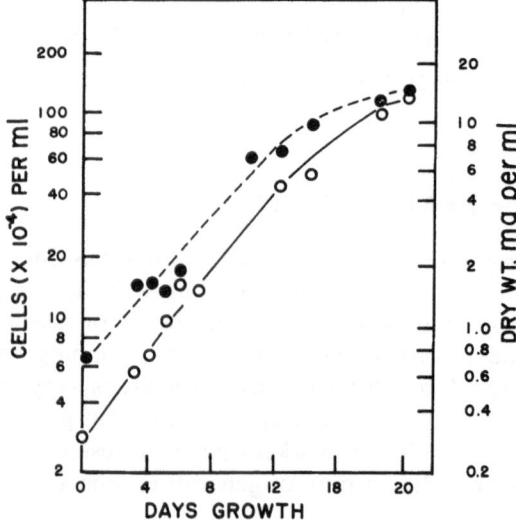

Fig. 5. Relationship between dry weight and cell number of cultured bean cells (Mandels *et al.*, 1968a). Murashige medium with 0.05 mg/l, 2, 4 D and 1.0 g/l phytone. ●– – –● Dry weight, ○———○ Cell count

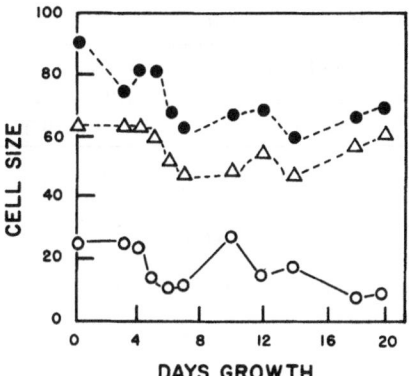

Fig. 6. Properties of bean cells in suspension culture (Mandels *et al.*, 1968a). Cells from Fig. 5. O————O Average cell dry weight, micrograms, ●– – –● Average cell length microns, △– – –△ Average cell width microns

ties of obtaining and diluting a representative sample. Optical density measurements are unsatisfactory (Dougall, 1966) because the plant cells are large (Fig. 6) and tend to grow in clumps so they do not form stable suspensions. Settled volume of tissue has been used as an approximate measure of growth (Graebe and Novelli, 1966a; Tulecke *et al.*, 1965). A few simple expressions of growth have been used such as the growth index which is the ratio of final to initial weight (Tulecke and Nickell, 1960; Wang and Staba, 1963); or productivity in mg per ml of culture per day (Mandels *et al.*, 1967; Miller *et al.*, 1968).

5. Large Scale Growth

Large quantities of plant cells of various species have been grown in suspension culture in several laboratories (Table 4,5). Although none of these studies have involved extensive investigations of bioengineering criteria they have given some insight into the special problems to be faced.

The media required may be expensive and difficult to obtain. Media used in the pioneering studies incorporated 5—15% coconut milk. This is the liquid endosperm that nourishes the coconut embryo. It is analogous to the sera used in animal cell culture. Other complex additives such as plant peptones or yeast extract may sometimes be substituted. More recent studies have used fully defined media with sucrose as carbon source.

A large inoculum (5—15% v/v) is required to shorten the lag phase and alleviate contamination problems. This can take weeks of scale up from

Table 4. Large scale growth of plant cell cultures

Culture (callus)	Vessel	Volume culture liters	Type	Productivity[a] g/liter fresh	per day dry	Ref.
Gingko pollen	Carboy	10	Batch	3.1	–	Tulecke et al. (1960)
Hollystem	Carboy	10	Batch	3.1	–	Tulecke et al. (1960)
Lolium endosperm	Carboy	10	Batch	3.1	–	Tulecke et al. (1960)
Rose stem	Pilot plant	134	Batch	9.7	–	Tulecke et al. (1960)
Carrot root	Flask	6	Batch	6.1	–	Byrne et al. (1962)
Spearmint stem	Carboy	3	Batch	4.7	–	Wang et al. (1963)
Rose stem	Phytostat	8	Semi-continuous	11	0.4	Tulecke et al. (1965)
Soybean root	Phytostat	2	Semi-continuous	–	1.3	Miller et al. (1968)
Corn endosperm	Flask	6	Semi-continuous	–	–	Graebe et al. (1966b)
Lettuce leaf	Fermenter	5	Semi-continuous	–	2.3	Mandels et al. (1968b)

$$^a Productivity = \frac{Total\ harvest - inoculum}{Culture\ volume \times days\ grown}$$

agar slant to shake flask to larger vessel. Use of established suspension cultures greatly simplifies preparation of inoculum.

Unlike bacteria, plant cells do not separate after cell division in the intact plant. This tendency persists in cell cultures and they may grow as large masses adhering to the baffles, above the liquid line, in the foam, or clogging air inlets, outlets, and harvest and feed lines. The use of established suspension cultures tends to give a more satisfactory type of growth. Friable fast-growing cultures are more satisfactory than slow-growing compact cultures.

Aeration is not a serious problem. Plant cells are obligate aerobes, but since growth rates are low, aeration rates are low and foam is usually light. Antifoam agents can be used if necessary and are usually non-toxic (Mandels et al., 1968a; Wang and Staba, 1963). The large thin-walled plant cells are easily broken so agitation must be gentle, although some agitation is essential if cultures are to grow (Graebe and Novelli, 1966a; Mandels et al., 1968a). In phytostats, cultures have been stirred by the aeration alone (Tulecke et al., 1965) or by intermittent use of a magnetic

Table 5. Growth of lettuce cells in semi-continuous culture (Mandels *et al.*, 1968a)

Experiment Number	1	2	3
Sucrose concentration (%)	3.0	3.0 – 2.0	2.0
Inoculum (grams)	10.0	4.0	9.7
Days operated	14	61	44
Average culture volume (liters)	4.9	7.1	6.1
Average feed rate (liters/day)	0.97	1.04	1.57
Harvest total (liters)	40.0	74.5	69.0
Ave. conc. grams (dry wt/liter)	12.0	9.5	8.1
Dilution rate per day[a]	0.20	0.15	0.26
Yield (grams/100 g sucrose)[b]	39	34	40
Productivity (g per liter per day)[c]	2.3	1.6	2.1

[a] $\dfrac{\text{Feed rate}}{\text{Ave volume}}$.

[b] $\dfrac{(\text{Total harvest grams}) - (\text{inoculum})}{(\text{Total liters}) \times (\text{g sucrose/liter})}$.

[c] $\dfrac{(\text{Total harvest grams}) - (\text{inoculum})}{(\text{No. days}) \times (\text{ave culture volume})}$.

Cells were grown on Murashige medium with 0.10 mg/liter naphthalene-acetic acid in a 15-liter New Brunswick Continuous Fermenter, Model CF 500, inoculum 1 liter suspension culture from shake flask, temp. $26 - 28°$ C, air $1 - 1.5$ liter/min, impeller 120 rpm. Exp. 1. Constant volume $1.5 - 2.5$ liter of culture harvested every $2 - 3$ days and replaced with fresh nutrient. Exp. 2, 3. Constant Nutrient Feed $1 - 1.5$ liter per day – Intermittent harvest.

stirring bar (Graebe and Novelli, 1966a). Conventional impellers at slow speeds have also been used (Mandels *et al.*, 1968a; Miller *et al.*, 1968).

Elaborate precautions must be taken to prevent contamination. Antibiotics such as tylosin, mycostatin (Byrne and Koch, 1962) amphotericin, griseofulvin, oxytetracycline and bacitracin (Wang and Staba, 1963) have been used by some workers without inhibiting growth of plant cells.

Because of the initial lag and the problems associated with getting a batch culture started, continuous cultures of plant cells offer marked advantages. Rose cells were grown (Tulecke *et al.*, 1965) in an 8-liter phytostat on Murashige medium. At intervals of one or two days about a liter of culture was harvested and an equal volume of fresh nutrient was added. In 7 runs totalling 232 days he harvested 163 liters of culture with an average fresh weight of 112 g (4.6 g dry) per liter or a productivity of 0.42 g dry weight per liter of culture per day at a dilution rate of 0.09 per day. The harvested cells averaged 3.4% dry weight, 16% of which was protein. The protein content ranged from 7% in slow-growing cultures to 19% in young cultures. Lettuce cells have been grown in semicontinuous culture (Mandels *et al.*, 1968a) in a New Brunswick laboratory fermenter either with intermittent harvest and addition of fresh nutrient (constant volume) or with continuous nutrient feed and intermittent harvest (Table 5). The harvested cells contained 8—19% protein. More elaborate devices for preparation of large quantities of tissue for biochemical studies (Graebe and Novelli, 1966 and 1966b) or for continuous culture and automatic sampling (Miller *et al.*, 1968) have also been described.

6. Final Remarks

As a fundamental science plant cell culture has made remarkable achievements and has greatly enriched our knowledge of plant growth and morphogenesis. As a bioengineering science it is in its infancy. The present status is that cultures can be obtained from almost any part of almost any plant. Many of these grow well in suspension on simple media. A few have been grown on a fairly large scale. Despite the obviously tremendous potential, fermentation studies with plant cell cultures have been few and limited. Animal cell culture work has received more attention because of medical applications and use in production of vaccines. The pioneering studies carried out with plant cells at Charles Pfizer and Co. (Tulecke and Nickell, 1959, 1960) ended when it became clear that costs would be great, and profitable fermentations too far in the future. The U.S. Army (Byrne and Koch, 1962; Mandels *et al.*, 1967; Mandels *et al.*, 1968a, 1968b) and the Air Forces (Tulecke *et al.*, 1965) supported some studies on the use of plant cells as a food or as a component of a bioregenerative system, but this work was also terminated for similar reasons and because no photosynthetic cultures were developed. In general the most serious problems are related to the slow growth of plant cells and problems of development and stabilization of high yielding strains. Because of these difficulties it would appear that future work should be directed towards production of high cost compounds such as drugs from rare or inaccessible plants.

References

Byrne, A. F., Koch, R. B.: Science **135**, 215 (1962).
Carew, D., Staba, E. J.: Lloydia **28**, 1 (1965).
Chan, W., Staba, E. J.: Lloydia **28**, 55 (1965).
Dougall, D. K.: Plant Physiol. **41**, 1411 (1966).
Earle, E. D., Torrey, J. G.: Plant Physiol. **40**, 520 (1965a).
Earle, E. D., Torrey, J. G.: Am. J. Botany **52**, 891 (1965b).

Gamborg, O. L., Eveleign, D. E.: Can. J. Biochem. **46**, 417 (1968a).
Gamborg, O. L., Miller, R. A., Ojima, K. O.: Exp. Cell. Res. **50**, 151 (1968b).
Gautheret, R. G.: The culture of plant tissues, p. 865. Paris: Masson et Cie. 1959.
Gibbs, J. L., Dougall, D. K.: Science **141**, 1059 (1963).
Graebe, J. E., Novelli, G. D.: Exp. Cell. Res. **41**, 509 (1966a).
Graebe, J. E., Novelli, G. D.: Exp. Cell. Res. **41**, 521 (1966b).
Halperin, W.: Science **146**, 406 (1964a).
Halperin, W., Wetherell, D. E.: Am. J. Botany **51**, 274 (1964b).
Hildebrandt, A. C.: Science **153**, 1287 (1966).
Hildebrandt, A. C.: In: Tracey, M. V., Linskens, H. F. (Eds.): Modern methods of plant analysis. Vol. V., p. 383. Berlin-Heidelberg-New York: Springer 1962.
Jones, L. E., Hildebrandt, A. C., Ricker, A. J., Wu, J. H.: Am. J. Botany **47**, 468 (1960).
Laetsch, W. M., Stetler, D. A.: Am. J. Botany **53**, 798 (1965).
Mandels, M., Maguire, A., El-Bisi, H. M.: Tech. Rep. 68-6-FL. U.S. Army Natick Laboratories, Natick, Mass., p. 56 (1970).
Mandels, M., Matthern, R. O., El-Bisi, H. M.: Tech. Rep. 69-22-FL. U.S. Army Natick Laboratories, Natick, Mass. p. 30 (1968a).
Mandels M., Jeffers, J., El-Bisi, H. M.: Tech. Rep. 69-36-FL. U.S. Army Natick Laboratories, Natick, Mass. p. 69 (1968b).
Miller, R. A., Shyluk, J. P., Gamborg, O. L., Kirkpatrick, J. W.: Science **159**, 540 (1968).
Morel, G.: Am. Orchid Soc. Bull. **33**, 473 (1964).
Muir, W. H., Hildebrandt, A. C., Riker, A. J.: Am. J. Botany **45**, 589 (1958).
Murashige, T., Skoog, F.: Physiol. Plantarum **15**, 475 (1962).
Nickell, L. G., Torrey, J. G.: Science **166**, 1068 (1969a).
Nickell, L. G., Maretzki, A.: Physiol. Plantarum **22**, 117 (1969b).
Quatrano, R.: Plant. Physiol. **43**, 2057 (1968).
Shimada, T., Sakakuma, T., Tsunewaki, K.: Can. J. Genet. Cytol. **11**, 294 (1969).
Sievert, R. C., Hildebrandt, A. C.: Am. J. Botany **52**, 742 (1965).
Steward, F. C., Mapes, M. O., Kent, A. E., Holsten, R. D.: Science **143**, 20 (1964).
Tulecke, W., Weinstein, L. H., Rutner, A., Laurencot, H. L., Jr.: Contrib. Boyce Thompson Inst. **21**, 291 (1962).
Tulecke, W., Nickell, L.: Science **130**, 863 (1959).
Tulecke, W., Nickell, L.: Trans. N. Y. Acad. Sci. **22**, 196 (1960).
Tulecke, W., Taggert, R., Colavito, L.: Contrib. Boyce Thompson Inst. **23**, 33 (1965).
Vasil, I. K., Hildebrandt, A. C., Riker, A. J.: Science **146**, 76 (1964).
Vasil, I. K., Hildebrandt, A. C.: Science **147**, 1454 (1965a).
Vasil, I. K., Hildebrandt, A. C.: Science **150**, 889 (1965b).
Wang, C., Staba, E. J.: J. Pharm. Sci. **52**, 1058 (1963).
Weinstein, L. H., Nickell, L. G., Laurencot, H. L., Jr., Tulecke, W.: Contrib. Boyce Thompson Inst. **20**, 239 (1959).
Weinstein, L. H., Tulecke, W., Nickel, L. G., Laurencot, H. J., Jr.: Contrib. Boyce Thompson Inst. **21**, 371 (1962).
White, P. R.: The cultivation of animal and plant cells, p. 246. New York: Ronal Press 1963.
Yatazawa, M., Furahashi, K., Suzuki, T.: Nippon Dojo-Hiryogaku Zasshi **38**, 305 (1967).

Mary Mandels, Ph. D.
Microbiology Division
Food Laboratory
US Army Natick Laboratories
Natick, MA 01760/USA